Sound Curriculum

Part of a growing group of works that addresses the burgeoning field of sound studies, this book attends not only to theoretical and empirical examinations, but also to methodological and philosophical considerations at the intersection of sound and education. Gershon theoretically advances the rapidly expanding field of sound studies and simultaneously deepens conceptualizations and educational understandings across the fields of curriculum studies and foundations of education. A feature of this work is the novel use of audio files aligned with the arguments within the book as well as the discussion and application of cutting-edge qualitative research methods.

Walter S. Gershon is Associate Professor in the School of Teaching, Learning & Curriculum Studies and LGBTQ Affiliate Faculty at Kent State University, USA.

Studies in Curriculum Theory
William F. Pinar, Series Editor

For additional information on titles in the Studies in Curriculum Theory series, visit www.routledge.com/education

Curriculum in a New Key
The Collected Works of Ted T. Aoki
Pinar/Irwin (Eds.)

Expanding Curriculum Theory
Dis/Positions and Lines of Flight
Reynolds/Webber (Eds.)

Schooling, The Puritan Imperative, and the Molding of an American National Identity
Education's "Errand Into the Wilderness"
McKnight

International Handbook of Curriculum Research
Pinar (Ed.)

Curriculum and the Holocaust
Competing Sites of Memory and Representation
Morris

Like Letters In Running Water
A Mythopoetics of Curriculum
Doll

Teaching as a Reflective Practice
The German Didaktic Tradition
Westbury/Hopmann/Riquarts (Eds.)

Curriculum as Institution and Practice
Essays in the Deliberative Tradition
Reid

Queer Theory in Education
Pinar (Ed.)

The Lure of the Transcendent
Collected Essays by Dwayne E. Huebner. Edited by Vikki Hillis. Collected and Introduced by William F. Pinar
Huebner

Sound Curriculum
Sonic Studies in Educational Theory, Method, & Pratice
Gershon

Sound Curriculum

Sonic Studies in Educational Theory, Method, & Practice

Walter S. Gershon

Routledge
Taylor & Francis Group
LONDON AND NEW YORK

First published 2018 by Routledge

2 Park Square, Milton Park, Abingdon, Oxfordshire OX14 4RN
52 Vanderbilt Avenue, New York, NY 10017

Routledge is an imprint of the Taylor & Francis Group, an informa business

First issued in paperback 2019

Library of Congress Cataloguing-in-Publication Data
A catalog record for this book has been requested

ISBN: 978-1-138-69209-1 (hbk)
ISBN: 978-0-367-34260-9 (pbk)

Typeset in Bembo
by Apex CoVantage, LLC

Contents

Preface

We live surrounded by sounds. They course through our bodies and resound across millennia, particles bombarding our soundfields in waves from distant galaxies and around the corner. From birth to death our lives are informed by sonic information, sound knowledge that helps us negotiate our everyday experiences. This is the case regardless of how much one hears, for sounds are felt as much they are aurally experienced.

Sounds are educational. From Ms. Clavel who wakes in the middle of the night in Ludwig Bemelmas' famous children's book *Madeline* to educators who have modulated their voices to reach students throughout the history human experience, such information is not necessarily news. Yet, more often than not, it is the ocular, our vision, that metaphorically and literally guides educational experiences. As a result, sound ideas often interrupt commonplace educational understandings and, in so doing, provide a means to consider everything from theories that strongly inform what educational dreams are possible to methodologies that examine everyday educational interactions to daily educational practices.

Not unlike the fields of sensory studies and affect theory, similarly interdisciplinary disciplines that share significant currents in its waters, what's new here is recognition of sound studies as a field of study in education, as well as a reorganization of its components. Similar to sensory studies and affect theory, sound studies is a large umbrella that houses scholarship from a wide array of academic disciplines that share a focus on the sonic. For example, Jonathan Sterne's (2012) edited work, *Sound Studies Reader*, includes contributions from such fields as anthropology, linguistics, media/communications, musicology, philosophy, and the sciences.

However, before this might be misinterpreted as an inadvertent argument that sound studies truly is a new field of study, it is important to note that there is a long history of attention to (the scholarship of) sound. This can be seen not only throughout Western constructions of meaning and quests for knowledge (e.g., Erlmann, 2010; Schwartz, 2011) but also across the breadth of the 20th century (e.g., Benjamin & Rosenthal, 2014; Russolo, 1967/1913) as well as in *musique concrete*, *avant garde* movements in Western art (classical) music, radio and sound productions, discussions of aesthetics, studies of music and people, and on and on. From this perspective, sound studies is at once a coalescing of multiple sound fields that are being brought together in relatively new and exciting ways and an expression of historical, deep connections to soundworks.

This combination of longstanding historical connections and recent expression is also reflected in educational sound studies. For example, R. Murray Schafer, the scholar/composer/artist who coined the term "soundscape," wrote *The Composer and the Classroom* (1965), four years before his work *The New Soundscape* (1969) and a decade before perhaps his most well-known work, *The Tuning of the World* (1977). In the U.S. canon of educators, Anna Julia Cooper (1892), W.E.B. DuBois (1903), and John Dewey (1980/1934), to name but three foundational scholars in U.S. education, all attend to questions of music and sounds as educational ways of being and knowing. More recently, scholars such as Ted T. Aoki (e.g., Pinar & Irwin, 2005), Liora Bresler (e.g., 1992, 2005, 2009), Greg Dimitriadis (e.g., 1996, 2009), Fredrick Erickson (e.g., 1982, 2003, 2004), and scholarship at the intersection of music, education, and curriculum (e.g., Emdin, 2010; Gaztambide-Fernandez, 2011; Ibrahim, 2014; Land & Stovall, 2009; Love, 2012; McCarthy, Hudak, Miklaucic, & Saukko, 1999), exemplify continued attention to the multiple intersections of sounds, music, and education. Nonetheless, although these works could, and I argue should, be understood as sound studies in education, there is little work that expressly claims the field of sound studies or seeks to utilize this growing set of sonic possibilities, either metaphorically or literally.

This collection of new and updated essays offers one set of possible trajectories for educational sound studies in theory, method, and practice. Due in no small part to an overemphasis on text and the ocular in contemporary educational thought as further detailed in the opening chapters of this book, attention to the sonic provides a strong set of discursive and material tools for listening to the flow of educational possibilities and otherwise interrupting commonsense notions of education and schooling.

These affectively vibrating beingknowingdoing waves produce resonances. Such resonances are imbricated across and within multiple simultaneous layers of beingknowingdoing[1]—literal, metaphorical, embodied. As further outlined in the chapters following this introduction, resonances are not only a means for considering sonic inquiry in qualitative research but also provide an educational philosophy through which the components of this book can be conceptualized. Where relevance is a consensus perspective, anything has the potential to resonate with any other thing, from the associated sound files, to the information expressed throughout this book, to others' ideas and ideals, to every aspect of local and less local ecologies, to those that affect you as you sit reading these words.

Like an open omnidirectional microphone, sound possibilities are overlapping and occur at multiple frequencies including those above and below perception that are nonetheless present. Unlike sight where a gaze frames parameters of the possible, as Seth Kim-Cohen (2009) argues, there is no blink of an ear. Otologically speaking, perception is more a question of filtering a vast simultaneity of information than focusing on a particular point or direction. This is not to say that listening/hearing somehow doesn't require a kind of focus or directionality or that eyes somehow don't filter. Instead, it is the qualitative difference between information one can't help but hear[2] as opposed to the ability to avert one's gaze or close one's eyes.

Chapters in this book echo the productively messy, necessarily unfinished, nested and layered nature of scholarship in the fields of curriculum and sound studies. Theories are presented alongside empirical examples, methods are theoretically grounded, and empirical studies loop back to the theoretical. The first three chapters provide an argument for what I call sound curriculum, articulate a position that sounds form educational systems of knowledge, and enunciate a philosophy of educational resonance through which the remainder of this book can be conceptualized.

Although there is an intentional flow to the remaining chapters in this book, chapters can be read in any order. For those wishing to move in a more linear and sequential fashion, the three initial chapters briefly described above are followed by an argument about policing sound in schools in which we wish hearing students deaf and deaf and hard of hearing students fully hearing, and both to only hear the teacher. The fourth chapter bridges the two overarching sections of the book, moving from theory to practice, introduces qualitative modes of sonic inquiry, describes some of the practical aspects of sound methodologies, and seeks to draw some distinctions between narrative radio and sound scholarship. The final three chapters that follow are more methodologically driven and address sound theory in education through empirical studies: one that speaks to the possibilities and pitfalls of applications for sound in daily classroom lessons, another that attends to questions of improvisation as practice and metaphor, and a final chapter that is a presentation of sound art and a discussion of how that work came to be.

Sound files appear throughout this book. For increased accessibility you can access these sounds in one of three ways: through a QR code at the beginning of each chapter that links to a webpage of the sounds throughout the chapter, by selecting a hyperlink in e-versions of this work, or by entering the following URL: www.soundcurriculum.net. If you are reading a traditional book, the place where sound files appear throughout is also the place you are intended to listen to and play the associated file on the website.

While I have worked to use text as clearly as possible, as this book argues, sometimes text, vision, and written words fall short. Sounds do indeed impact listeners in ways that are qualitatively different than text and images (see Chapters 1–3) and those associated with the various points raised throughout this book have been intentionally included in order to make or explicate a given argument. In other words, although one might understand the points made, understandings here are incomplete without the sonic.

This, then, is the central message of this book: Sounds are educational and can be used to great advantage to consider and reconceptualize what education might mean and how those understandings could function in practice. It is an opening of possibilities rather than an exhaustive list or a set of concrete directions that should be followed. In previous writings, I have been perhaps not as direct as I might have been about my intentions as I land theory in everyday experiences or note potential theoretical consequences for methodological decisions. To be clear, my purpose in this preface and throughout this book is to articulate the flexibility and practicality for possibilities that can be organized as a field of educational sound studies. Examples are precisely that, a particular iteration of sound ideas and sonic affects that combine to underscore both an argument for educational sound studies and the waves of scholarship that might traverse this field, resonances I call sound curriculum.

Notes

1 Although there is more talk in Chapter 3 about this construction, it is in many ways as it seems, an effort to enunciate a lack of separation between being, knowing, and doing in ways that are still often nonetheless separated, as onto-epistemology is still separated by a hyphen that insists on deeper distinctions between ontology and epistemology than may often exist, for example. It is at the same time an effort to explicitly err in the opposite direction, to suggest that being and knowing and doing are all inexorably imbricated in ways that their articulation in this fashion emphasizes the consistency of their relations rather than the exceptions of their separation.

2 This is the case for hearing people. For the deaf and hard of hearing, it's often information that one can't help but to feel, a point I address in the following chapter on sounds as educational systems of being and knowing.

References

Benjamin, W., & Rosenthal, L. (Ed.). (2014). *Radio Benjamin*. New York: Verso.

Bresler, L. (1992). Qualitative paradigms in music education research. *The Quarterly Journal of Music Teaching and Learning, 3*(1), 64–79.

Bresler, L. (2005). What musicianship can teach educational research. *Music Education Research*, 7(2), 169–183.

Bresler, L. (2009). Research education shaped by musical sensibilities. *Cambridge University Press: British Journal Music Education, 26*(1), 7–25 doi:10.1017/S02650517080082

Cooper, A. C. (1892). *A voice for the South by a Black woman from the South*. Xenia, OH: Aldine Printing House.

Dimitriadis, G. (2009). *Performing identity/performing culture: Hip hop as text, pedagogy, and lived practice* (Revised ed.). New York: Peter Lang.

DuBois, W. E. B. (1903). *The souls of Black folk*. Chicago: McClurgand.

Emdin, C. (2010). *Urban science education for the hip-hop generation*. Rotterdam, NL: Sense Publishers.

Erickson, F. (1982). Classroom discourse as improvisation: Relationships between academic task structure and social participation structure in lessons. In L. C. Wilkinson (Ed.), *Communicating in the classroom* (pp. 153–181). New York: Academic Press.

Erickson, F. (2003). Some notes on musicality and speech. In D. Tannen & J. E. Atalis (Eds.), *Linguistics, language and the real world: Discourse and beyond (Georgetown University: Round table on languages and linguistics)* (pp. 11–35). Washington, D.C.: Georgetown University Press.

Erickson, F. (2004). *Talk and social theory: Ecologies of speaking and learning in everyday life.* Malden, MA: Polity Press.

Erlmann, V. (2010). *Reason and tesonance: A history of modern aurality.* Cambridge, MA: Zone Books.

Gaztambide-Fernandez, R. (2011). Musicking in the city: Reconceptualizing urban music education as cultural practice. *Action, Criticism & Theory for Music Education, 10*(1), 15–46.

Ibrahim, A. (2014). *The rhizome of Blackness: A critical ethnography of hip-hop culture, language, identity, and the politics of becoming.* New York: Peter Lang.

Kim-Cohen, S. (2009). *In the blink of an ear: Toward a non-cochlear sound art.* New York: Continuum.

Land, R. R., & Stovall, D. O. (Eds.). (2009). Hip hop and social justice. [Special Edition] *Equity and Excellence in Education,* [Special Issue]. *42*(1).

Love, B. (2012). *Hip hop's li'l sistas speak: Negotiating hip hop identities and politics in the New South.* New York: Peter Lang.

McCarthy, C., Hudak, G., Miklaucic, S., & Saukko, P. (Eds.). (1999). *Sound Identities: Popular music and the cultural politics of education.* New York: Peter Lang.

Pinar, W. F., & Irwin, R. L. (2005). *Curriculum in a new key: The collected works of Ted T. Aoki.* New York:R outledge.

Russolo, L. (1967/1913). *The art of noise* (R. Filliou, trans). New York: Something Else Press.

Schafer, R. M. (1965). *The composer in the classroom.* Scarborough, ON, Canada: Berandol Music Limited.

Schafer, R. M. (1969). *The new soundscape: A handbook for the modern music teacher.* Scarborough, ON, Canada: Berandol Music Limited.

Schafer, R. M. (1977). *The tuning of the world.* New York: Knopf.

Schwartz, H. (2011). *Making noise: From Babel to the big bang and beyond.* Cambridge, MA: Zone Books.

Sterne, J. (Ed.). (2012). *The sound studies reader.* New York: Routledge.

Acknowledgements

There are so very many to thank as the ideas that inform this book took shape over the past decade and a half. When I first waded into intersections of education and sound, I was perhaps most interested in how music, both as metaphor and lived practices, might serve as a critically creative tool for considering educational interactions. More specifically, I was thinking about forms and possibilities for improvisation. It was through this work that I found that just about everything I'd ever considered at the time was something Fredrick Erickson had not only considered but already written about in his usual complexly clear fashion, often decades earlier. This was also my experience in discovering the work of Ted T. Aoki, a scholar whose attentions to the complex interrelations between curriculum, sound, and wonder paved the way for my own. In my second year as a doctoral student, I gained an unexpected ally in a person who became a sound-touchstone in my life until his untimely death, Greg Dimitriadis. Then there is the inimitable Liora Bresler whose work in sound, music, and qualitative research, and education in general, I've read since my earliest days as a graduate student and have had the pleasure of thinking with on and off over the past six or seven years. I would be utterly remiss if I didn't thank Bill Pinar, who has been an unwavering ally, mentor, and friend and in whose series this book has been published.

I would also like to thank my sonic traveling companions who have been my sounding board as we traversed possibilities across fields and disciplines, people who always have smart things to say and are willing to lend an ear: Peter Appelbaum, Regan Mitchell, Kim Powell, Sam Rocha, and Boni Wozolek. Their collective and individual input has strengthened and deepened this book in ways that would have been impossible without their time, effort, and care. Here, Boni Wozolek gets another shout out for lending her considerable critical understandings to earlier drafts of chapters on method and songwriting, as does Reagan Mitchell for reading and then pressing at my thoughts at the intersection of sound, education, and philosophy.

Thanks also go to Joe Valente on matters of critical deaf and disabilities studies; Mirka Koro-Ljunberg and the doctoral students in her Data and Data Construction in Qualitative Research class who read and commented on a draft of Chapter 3 as part of our Skype conversation this past (2016) fall; and Philip Gelb for his input on my educational translation of his friend, mentor, and musical collaborator Pauline Oliveros's work on deep listening. Thank you also to all the musical mentors—including Maura Bank, Luca Brandoni, Angel Luis Figueroa, Munyungo Jackson, Badal Roy, V. R. Venkataraman, and John B. Williams—who taught me not only how to be a better player but also how to listen. It may have been a minute since we've been in each other's orbits but your lessons keep on giving.

Thank you to John Bennett, Karen Grindall, Pat Marxen, Dolores Whaley, and all the first, fifth, seventh, and eighth grade students whose work with me in the Listening to the Sounds of Science Project in many ways triggered and drove my deepening understandings of what I now call sound curriculum. Thank you to the anonymous reviewers whose thoughtful suggestions and positive recommendations made this work a possibility. Thank you to Naomi Silverman for your patience as I strove to get this book in publishable shape and to the rest of the team at Routledge who helped turn this manuscript into a book. All (inevitable) mishearings are my own and in spite of everyone's not insignificant efforts to the contrary.

Thank you to Dena and Kate Gershon for your love and support, this is what I was working on all that time.

1 Sound Curriculum

Breaking Frames and Opening Ears

Introduction

This introductory chapter sits at the intersection of questions about curriculum, methodology, perception, and representation. As I argue below, understanding knowledge in, as, and through sound maintains current curricular ways of knowing while opening new doors for conceptualizing and interpreting curriculum. By doing so, I take seriously Stoller's (1997) call for "sensuous scholarship" and Ellsworth's (2005) notion that all understandings are a literal "making sense"—knowledge gained through one's senses (for example, Feld and Basso, 1996; Gershon, in press a; Howes, 1991, 2005).

Listening brings to the fore not only the importance of hearing and the senses in schooling, but also the central nature of *all* sounds in educational contexts. Because my goal here is to articulate what might be gained through sonic inquiries and representations of educational contexts, this discussion is not intended to be an exhaustive review of either how sound inquiry is approached across disciplines or the myriad ways in which such uses might be applied to curriculum and related fields of study. Rather, my purpose here is twofold, to provide some criteria for recognizing and conceptualizing sound curriculum as it currently exists in the field and to present how the further conceptualization of sound curricular possibilities can contribute to advancement in the field of curriculum studies.

Understanding the existence and potential of sound curriculum is simultaneously a question of framing, a movement beyond the frame, and a critique of framing. After a brief discussion of framing that describes this trajectory, my argument here is comprised of three central components. It begins with an act of recognition, framing previously disparate examinations of music, sound, and meaning in curriculum studies as sound curriculum. I then draw from the burgeoning field of sound studies in order to complicate the sound curricular understandings presented in the first section, a breaking of the frame to note its boundaries. This then leads to questioning the meaning and limitations of the verb *to frame*. A final brief conclusion suggests some possible directions for sound curriculum and how it might contribute to the complicated conversations that are contemporary curriculum studies.

I have elected here to begin with Judith Butler's (2009) discussion of framing, one of the more sophisticated, fluid discussions of frames and their applicability for attending to notions of equity, access, and life. When rewriting this piece so that it served as the introduction to this book, I elected to keep this discussion of Butler and framing for the following reasons. First, because it is a comparatively recent, deeply nuanced, and highly sophisticated discussion of framing, Butler's arguments press at many previous constructions that address what framing is and how it can function. Second, as Butler's talk about framing in this instance addresses her concerns with how people can be framed as less than human in order that they might be violently and aggressively attacked (murder, genocide), Butler's presentation of framing is decidedly ethical in nature and, as such, attends to issues of social justice that are central to my project here. Finally, Butler's scholarship here is masterful, articulating fluid frames that drop, are made, and again reframed in multiple layers that move away from the original framing in ways that are all the more meaningful in spite of their movement and mutation.

And yet, Butler, whose conceptualization of frames and framing encapsulates and pushes past many previous constructions in ways that are, as noted here, more ethical, fluid, and complex, is still limited by the ocular baggage of her argument. It is for these and other such reasons that I begin here with Judith Butler's arguments about frame and framing as a springboard for discussions about the possibilities of sound and the power of the sonic in considering educational ways of being and knowing.

However, a central question remains: What is sound curriculum? One answer was provided at the end of the preface, the wide variety of attention to the sonic in educational scholarship and the resonances within, between, and outside of those understandings. From this perspective, a key purpose of this book is to provide an open, fluid foundation for what might be conceptualized as "sound curriculum." Another answer also present throughout this book is less academically oriented: how anything that might be conceptualized as sonic contributes to our understandings and, in so doing, is sound knowledge, the topic of the following chapter. A third possible definition, and one that in many ways encapsulates the previous two, is sound ways of beingknowingdoing. By this I mean the myriad ways the sonic, sound metaphors, sound practices, sound ideas and ideals, and sound ways inform how we "is," what we "be," and that which we "know."

My resistance to landing on any one definition is that the sonic has so many possible interpretations and pathways that emphasizing any singular direction feels as misguided as trying to pin a sound down with a stapler. And, oxymoronically, not providing some sort of definition feels at once irresponsible to a project about something as central to experience as the sonic and runs the risk of being misunderstood or misconstrued. Given that this conundrum lies at the base of many attempts to graphically express the sonic (e.g., Sauer, 2009), this book is to sound curriculum as all forms of musical notation are to the expression of music: representations that are not to be mistaken as sounds, how those sounds affect things and ecologies, or the ways in which the sonic is expressed or interpreted, all of which are fluid, omnidirectional, and ever-present, a messy, interwoven complexity that in no way removes its sensation, signification, or significance.

Of Butler and Frames

Judith Butler (2009) contends the frames that one utilizes to conceptualize another person's life as being of value are socially constructed, predicated on existing norms of recognizability and "schemas of intelligibility [that] condition and produce norms of recognizability" (p. 7). For Butler, these frames derive their power as they move through successive iterations of framing and reframing towards intents and purposes different than those conceptualized by the original frame(rs). While such framing often serves to propagate dominant norms and values, it is in their reframing that such ideas and ideals can be subverted or interrupted. In short, frames demarcate how life is recognizable and what can be valued, structures that gain their power and meaning as they break from themselves.

> As a consequence, the frames that, in effect, decide which lives will be recognizable as lives and which will not, must circulate in order to establish their hegemony. This circulation brings out or, rather *is* the iterable structure of the frame.
>
> (Butler, 2009, p. 12, emphasis in original)

On one hand, Butler's notions of framing can be understood as moving beyond many previous notions of framing that, while certainly useful in considering how people categorize their understandings, are nevertheless often structuralist and universalist in construction. Gregory Bateson's (1972) often-cited framing of play is but one example of this tendency. He argues that play is universally understood and exists as a space within clear boundaries between the imaginary and reality, lines that clearly indicate what is play and what is not, similar to the way a frame delineates the artwork it contains from the non-art outside its boundaries. Both Vygotsky (1976) and Piaget (1962) similarly envisioned play as a universally applicable, non-serious space for children to try on roles, interactions, and ideas as well as a space to explore the imaginary. These understandings of play are germane to the points I wish to make because they speak to issues of frames and framing, as does more recent scholarship that challenges the kinds of static, universally understood frames of play that Bateson (and by extension Vygotsky and Piaget) propose.

Scholarship critical of universal claims of play often note that there are ethical implications in this kind of static, universal framing—what one person in a given interaction considers play can be threatening to another player (for example, Houseman, 2001; Lindquist, 2001). In lieu of framing play, where Borges (1998) posits that different kinds of play interactions reside alongside one another like a garden, Handelman (2001) suggests that the ethical, socially constructed nature of play is more akin to a Möbius strip, constantly folding in on itself, making and remaking itself as it moves. The advantages each of these constructions of play have over Bateson, Piaget, and Vygotksy's non-dynamic, universally applicable, and binary-creating frames—play as interactions that are universally understood, constant, and that reside outside of "seriousness" of the "real world"—is in how metaphors of gardens and Möbius strips expressly note the complex relationship between ethics, sociocultural norms and values, and epistemological categorization.

Although Butler (2009) retains the frame as a central construct, her discussion of the ways a frame breaks on each iterative framing and links the ability to frame anything to those who are entitled to make such decisions through their sociocultural/economic/historical capital, does indeed seem to represent a philosophical evolution in the consideration of frames as an epistemological, ontological construction. Although she does not make her argument in reference to their work, through this complexifying of previously static notions of framing, Butler has managed to capture both the iteratively recursive nature of Handelman's Möbius strip and the parallel motion in Borges' garden metaphor.

On the other hand, Butler's discussion of framing retains some ironically Western and structuralist metaphorical assumptions. This can be seen in the duality of her insistence that non-visual framings are as important as visual framing (texted, seen). While her inclusion of non-visual ways of knowing is certainly to be applauded as it is a step that is still often overlooked, this position nevertheless carries the following assumptions about the nature of visual metaphors, constructions that seem to imply a continuing Western, structuralist understanding of sense-making.

First, there is the notion of a frame, a visual metaphor. How does a frame smell? Or, more germane to the discussion in this chapter, what does a frame sound like? Second, Butler's suggestion that frames can be applied to other sensory ways of knowing implies that visual metaphors are equally applicable to non-visual senses. Not only has this been demonstrated not to be the case in both Western and non-Western constructions of sense-making (e.g., Erlmann, 2010; Stoller, 1989) it is also an indication of the primacy of the eye (Aoki, 1991; Ihde, 2007; O'Callaghan, 2007). Finally, when placed in combination, such understandings appear to note a rather Western construction of meaning that retains a Cartesian separation between the senses and their limitation to five discreet senses, an understanding that often serves to blur embodied knowing inasmuch as it helps explicate sensory knowledge and expressions of being (Howes, 2003; Stoller, 1997).

This does not mean, however, that Butler's construction of framing is not useful, nor do I mean to incidentally imply that more static conceptualizations or uses of frames are somehow passé. Rather, I seek to enunciate these differences as I travel through them for the remainder of this chapter. First using a static notion of framing to underscore the presence of sound curriculum in the field, then breaking that frame in order to expand the notion of sound curriculum beyond its expression as talk and music, and then finally suggesting the possible merits of further exploration of sound curriculum that exceeds notions of framing as an apt metaphor for sound understandings.

Framing Sound Curriculum: Understanding Curriculum as Sound Texts

Although it has not been recognized in this fashion, there is history of scholarship in the field of curriculum studies that can be framed as "sound curriculum." Sound curriculum is evident not only in aesthetic and artistic discussions of knowledge and meaning-making but also in relatively commonplace examinations that regard both the daily life of schools and the norms and values that contextualize those daily interactions. Specifically, sound curriculum is most often evident as talk and, in decreasing frequency, as music and sound.

Sound Curriculum, Organized and Texted

In curriculum studies, sound is nearly always organized and texted. There is a continuing debate over the definitions of how the organization of sounds is most clearly enunciated and the ramifications of those definitional categorizations. I use the terms "organized" and "sound" here to refer to the intentional ordering of pitches, tones, and rhythms that are most often manifest as what is more conventionally referred to as "music" and "talk." By texted I am referring to both the action and the idea that aurally perceived events can be represented notationally as some form of text, from the written word to graphic notations of musical ideas (for example, Oliveros, 2013; Sauer, 2009).

This tendency is due in no small part to the ways in which curriculum is often conceptualized. Practically, there are difficulties in representing the kinaesthetic and aural aspects of dance and music as text, be they virtual or physical. However, conversely, there is an emphasis on curriculum-as-text (e.g., Morris, 1999; Pinar, Reynolds, Slattery, & Taubman, 1995) that also often surfaces in studies at the intersection of the arts and curriculum (e.g., Dimitriadis, 2009; Springgay, 2002). Such framing is not surprising as it reflects the "literary turn" in the social sciences (e.g., post-colonial, post-modern, and post-structural discourses) that construct contexts and human interactions as "big T" Texts and little "t" texts. This allows ideas and actions to be deconstructed and otherwise critically examined the same way in which one approaches literal texts, a possibility as common in aesthetic and artistic curriculum as it is in the rest of the field.

Marla Morris' (1999) *Curriculum as Musical Text* is a case in point. Morris writes eloquently of the interwoven layers of her musical past, curriculum present and future, and the modes that serve as the structures she negotiates as she makes explicit the notes she sounds through her improvising lines of connectivity. She does so by writing about music as curriculum as text: music as Western art music and constructions of musicianship within a comparatively static Western art music context. This combination invokes structuralist understandings of music that can be understood as antithetical to the rich narrative she weaves throughout the piece and the kinds of fluid postmodern constructions of meaning she evokes in her work (e.g., Dell'Antonio, 2004).

As with all the points central to my argument throughout this chapter, I am not arguing against constructing ideas and interactions as Text, as this allows the scholarly space for many powerful literary tools that can make oft-hidden habits of mind and practice explicit for examination and critique (e.g., Pinar, Reynolds, Slattery, & Taubman, 1995; Malewski, 2009; Tarc, 2015). However, privileging the visual can serve to miss important sensory knowledge and sound understandings (check, for example, Appelbaum, 2007, and Pinar & Irwin [Aoki], 2005). In addition, and somewhat ironically, post-modern, post-structural discourses about embodied, sensual understandings are often presented in relatively static, disembodied ways (Stoller, 1997), a tendency Les Back (2007) characterizes about his home field as an unintended consequence of sociologists' love of language: "The problem is that sometimes this love of language leads us to try to create one of our own and as a result weakens our ability as communicators" (p. 2).

Yet talk is a key facet in understanding curriculum. This is not only the case in interpretive studies of educational contexts (e.g., Spindler & Hammond, 2006; Valli & Chambliss, 2007), but is so ubiquitous that it is literally impossible to consider the field after the mid-1970s without a centrality of talk. However, it is the rare instance when talk is conceptualized as a kind of organized sound rather than as a mode of communication. In those relatively rare instances when talk in educative contexts is considered as sound, it is most often done so to illustrate the musical nature of that talk (e.g., Erickson, 1982, 2003; Chapter 5).

Music and Curriculum

More common in curriculum studies are discussions of music as a springboard for curricular explorations and discussions of music as a curricular tool or cultural bridge. The following two works are strong cases for the historical significance of sound in curriculum studies, Anna Julia Cooper's (1892) *A Voice From the South, By a Black Woman From the South* and W.E.B. DuBois (1903) *The Souls of Black Folks*. Cooper frames her entire book around two musical phrases: soprano *obligato* and *tutti ad libitum*. *Obligato* [Obbligato] is a line of music that is to be played as written, an integral, significant, and essential aspect of the score. *Ad libitum* is often now expressed as "ad-lib," meaning expressed as one sees fit with less to no obligation to the score. As most know, soprano is one of the two categories for female voices in Western art music and *tutti* means as a group, together, all, pulling out all the stops on the organ. Cooper's musical instructions are at once a call for her to be given the trust and dignity of intentional meanings, and to work together so that all voices and processes are possible as the individuals see fit, a point to which I will return in the third chapter on resonance. Music is similarly central to DuBois, in both this and other works (e.g., DuBois, 1926), and I quote him here at length:

> What are these songs, and what do they mean? I know little of music and can say nothing in technical phrase, but I know something of men [sic], and knowing them, I know that these songs are the articulate message of the slave to the world. They tell us in these eager days that life was joyous to the black slave, careless and happy. I can easily believe this of some, of many. But not all the past South, though it rose from the dead, can gainsay the heart-touching witness of these songs. They are the music of an unhappy people, of the children of disappointment; they tell of death and suffering and unvoiced longing toward a truer world, of misty wanderings and hidden ways. The songs are indeed the siftings of centuries; the music is far more ancient than the words, and in it we can trace here and there signs of development.
>
> (DuBois, 1903, n.p., retrieved from www.gutenberg.org/files/408/408-h/408-h.htm)

Woodson is having something of a long overdue resurgence of attention, DuBois is often the token African American scholar noted in historical texts of the field, and Cooper is similarly gaining attention at long last, due in no small part to Melissa Harris Perry's national attention and creation of the Anna Julia Cooper Center and Project (www.cooperproject.org) (e.g., Au, Brown & Calderón, 2016; Grant, Brown, & Brown, 2015). Yet, as has been argued recently elsewhere (e.g., Gershon, 2016; Wozolek, 2015), a general lack of attention to the depth and possibilities of what might be called African American intellectual traditions in the field of curriculum studies has also inadvertently added to Western, Eurocentric understandings of the sensory.

This is not an ad hominem, ahistorical sideswipe. There has been and continues to be attention to this tradition in curriculum studies (e.g., Castenell & Pinar, 1993; Watkins, 2001), not the least of which is Carl A. Grant, Keffrelyn D. Brown, and Anthony L. Brown's (2015) recently published work *Black Intellectual Thought in Education: The Missing Traditions of Anna Julia Cooper, Cater G. Woodson, and Alain LeRoy Locke*. However, a continuing attention to Continental philosophical traditions and emphasis on European schools of critical theory (e.g., German and British schools with Gramsci somewhere in the middle) often incidentally serves to miss the depth and breadth of critical scholars of color and women in the United States. Like curriculum studies, cultural studies too has a tendency to explore more contemporary expressions of people of African descent at the expense of its historical texts, a particular irony when Stuart Hall is one of your field's founding scholars. Again, this is a broad brush against which many counterexamples certainly exist. Yet, across the majority of critical traditions, it is more likely one finds Margaret Sanger than Anna Julia Cooper or Franz Fannon than Carter G. Woodson. And in so doing, various scholarly traditions have often missed the significance of sound, music, and song as central to the human experience, critical theorizing, and material consequences. This is particularly ironic when the sound-as-arts (e.g., music) are utilized to enunciate points about the liberatory possibilities of and for the arts (for more on this point, see Gershon, 2014b).

Within the field of curriculum studies, Maxine Green's (1995) use of music and art to consider questions of ethics in teaching and Jose Rosario's (1991) discussion of the how his encounters with John Coltrane opened his ears and his mind to ideas are but two cogent examples of how curriculum scholars use music as a springboard for theoretical exploration. My framing of music as a springboard for theoretical exploration also includes how scholarship uses musical metaphors, teacher-as-conductor for example.

In many ways following the path set out by Cooper and DuBois, works such as *Performing Identity/Performing Culture: Hip Hop as Text, Pedagogy, and Lived Practice* (Dimitriadis, 2009), *Sound identities: Popular music and the cultural politics of education* (McCarthy, Hudak, Miklaucic, & Saukko, 1999), Christopher Emdin's (2010) *Urban Science Education for the Hip-Hop Generation: Essential Tools for the Urban Science Educator and Researcher*, and Land & Stovall's (2009) special issue of *Equity and Excellence in Education* on "Hip Hop as Social Justice" are but three examples of how music has been utilized as a curricular tool or cultural bridge to better reach often marginalized individuals and/or groups. Similarly, where scholarship that considers processes of music-making as a curricular tool for understanding non-musical academic content is rare (e.g., Gershon, 2010), work that adopts musical structures is even rarer still (e.g., Gouzouasis, 2008).

Perhaps due to its location, there is little work in the field of curriculum studies that explores formal music education, an area of interest that tends to reside in the neighboring field of music education. Exceptions to this include Gustafson's (2008, 2009) discussions of the normalizing, marginalizing, and raced nature of music education in United States classrooms. Ruben Gaztambide-Fernández's work similarly challenges both "the myth of the musician as the locus of music making" (2010, p. 66) and emphasizes the need for urban music educators to work "both *with* and *against* the prevailing narrow conception of the 'urban' that shapes the way we think about both urban music and urban education" (2011, pp. 17–18). My own work has similarly challenged dominant norms and values about the nature of music as knowledge and the importance of sensuous, embodied understandings (see Chapter 2).

Uses of music in curriculum studies, however, are comparatively infrequent when placed alongside discussions of visual arts, the use of visual arts as bridges to curricular understandings, and literature and literary tropes in the field (e.g., Eisner, 1985, 2002; Pendergast, Leggo, & Sameshima, 2009; Springgay, Irwin, Leggo, & Gozouasis, 2007). As with an emphasis on Text, this can also be understood as an ironic unintended consequence of privileging the visual in the field. Parallel to post-studies that write of embodied experience in disembodied ways, curriculum scholars often critique dominant norms and values while reinscribing the Western primacy of the visual (for non-Western sensual understandings, see for example Feld, 1996, 2005; Howes, 2005; Stoller, 1997). Furthermore, mirroring studies of talk in curriculum studies, discussions of music in/as/through curriculum rarely conceptualize music as sound.

Therefore, in spite of the ubiquity of sound in people's daily lives (LaBelle, 2010; Sterne, 2012), it is relatively rare for sounds to be considered as *sounds* rather than as literal or metaphorical conceptualizations of talk or music. Additionally, in spite of being the most often examined iterations of the sonic in curriculum studies, both talk and music are very rarely considered as sound, the "media" (for lack of a better term) in which it is contained. Instead, it is rendered, manipulated, and explored through the visual medium of text (Aoki, 1991).

Sound and Curriculum

There is, however, a small set of curriculum scholarship that that is not organized as music or talk but rather on sounds, sonority, or silence (e.g., Pinar & Irwin [Aoki], 2005; Jardine, 2004/1992; Miller, 2005). The majority of these works focus on questions of silence (e.g., Delpit, 1995; Thornton, 2004; Weis & Fine, 2005). Emblematic of such work is Janet Miller's (2005) now widely read "The Sounds of Silence Breaking: Feminist Pedagogy and Curriculum Theory." For Miller, silence and quiet is an oppressive force, an act of silence that she fights to break by giving voice to questions and concerns that marginalize and silence women and feminist perspectives. "The fear is that unnatural silences will envelop me, either lulling me into comforting quiet of supposed completeness, or muffling the sound of my questions (Miller, 2005, p. 62).

Conversely, David Jardine's (2004/1992) "A Bell Ringing in the Empty Sky" is simultaneously a critique of "an unearthly discourse that we must always have something new and notable to say at a moment's notice" (p. 263) and a call to recognize "both the exquisite solitariness of nature, and the vibrant and crawling interrelatedness and resonances of the earth" (p. 264). In sum, Jardine is arguing that through an understanding of the interconnectedness of things, a relation is created to "those archaic sounds and resonances that were not from our own assembly in the first place. What emerges out from under the exhaustion is a deep love and compassion for all things" (p. 274).

The curriculum scholar who has perhaps most often and directly spoken to the relationship between sound and curriculum is Ted T. Aoki. From his discussions on the importance of listening to the "layered voices of teachers" (pp. 187–198), to the section that editors William F. Pinar and Rita L. Irwin call "sounds of curriculum spaces" (pp. 333–432), Aoki's commitment to and theorizing of sound curriculum is readily apparent in his collected works (Pinar & Irwin [Aoki], 2005).

Resonating with Jardine (who cites Aoki in his work) is Aoki's invited talk to the International Conference on Korean Studies, in which he "savor[s] the reading of this story that may allow for me a listening to the sound of pedagogy emerging from the earthly silence in the land of the morning calm" (Pinar & Irwin, p. 389). He also literally and figuratively "notes" the importance of hearing and listening as knowledge contained in the sounds of talk, music, and nature as well as the ways in which music is curriculum for those who listen with open ears. It is Aoki (1991) who explicitly reminds us that, "the primacy of the eye in curriculum talk" (p. 182) can serve to miss important sensory knowledge and sound understandings.

Further, returning to the point raised above about what Wozolek (2015) has framed as "presences and absences" within the field of curriculum studies, the presence of this strong work on sounds in curriculum studies also notes an incidental ahistorical absence of voices like Cooper's and DuBois'. For, as I have used in arguments along these lines elsewhere, it was DuBois (1926) who proclaimed that he does "not care a damn for any art that is not used for propaganda. But I do care when propaganda is confined to one side while the other is stripped and silent" (n.p.). In these ways, sound curriculum provides a means to both echo longstanding resonances across the imbricated constellation of trajectories that is contemporary curriculum studies (and studies of education more broadly) and to interrupt such discourses in critically affirming ways.

On Noise and Educational Contexts

Fields outside of curriculum studies have paid particular attention to sounds in schools and classrooms. For example, there has been growing interest from scholars in psychology, environmental studies, acoustics, and architecture on how sounds impact children in classrooms and schools. Although such studies often differ in their aims and methods, their examinations of the relationship between sound and education most often categorize sounds as unwanted noise.

This construction echoes Jardine's (2004/1992) characterization of the sounds of modern life: "the clangor in which many of us live—the noisy abode out of which our curriculum theorizing often arises as a hurried response to our shortened breath" (pp. 162–163). Given this orientation to the sonic, it is therefore not surprising that these and fields attending to the sonic in education find sounds to be detrimental to teaching and learning. This is particularly the case when either a) the students in question have difficulty hearing or some other way of knowing that makes concentration in non-interrupted settings difficult, or b) in urban contexts in which there is likely to be a greater influx of the sounds associated with cities entering the classroom (e.g., Serra & Biasooni, 1998; Shield & Dockrell, 2003), points taken up again in the following two chapters.

In sum, there appears to be two dichotomous pairs that govern sound in curriculum studies: silence vs. sound, sound vs. noise. Silence can be an oppressive force that needs to be shattered to give voice through sound or is sought in the face of noise as a place of respite and reflection. Conversely, sound can be understood to convey knowledge, either as a kind of curriculum and/or as a form of pedagogy, or is the unwanted noise that should be filtered out. In this construction, while both silence and sound can serve as positive sources of knowledge, only noise appears to have no place in curriculum studies.

Framing Sound Curriculum, Breaking the Frame

There is work in curriculum studies that has examined various aspects of sound as separate entities, as talk, music, silence, sound, and noise. Yet, although there has been a good deal of scholarship that considers talk and, to a somewhat lesser degree, music, there is relatively little scholarship that examines talk and music as sound. Nonetheless, as this brief recounting documents? there is a history in the field of work that addresses various aspects of the relationship between sound and curriculum. An act of recognition is required in order to understand these sometimes disparate-seeming works as sound curriculum requires and as Butler (2009) suggests. In the West, sound is that which can be perceived, recognized, or apprehended aurally through one's ears. Sound curriculum breaks these works' original framing as "music" or "talk" and recognizes them as iterations of curricular explorations of sound.

Of at least equal importance, even from a relatively limited Western five-senses model of the sensorium (Gershon, 2011), sounds are not merely aural. Sounds are haptic, touching our bodies in ways we cannot help but to feel—hearing sounds is itself a haptic experience, the pressing of airwaves against a vibrating tympan. It is also always the sonic waves that hit our bodies regardless of whether they are noticeably felt or exist within the range of human perception. As a deaf colleague recently shared with me, one of his more profound experiences after going deaf in his late teens involved being picked up by a deaf friend and colleague at the airport. After a short discussion in which they found they both had a love for classic rock in general and the Pink Floyd in specific, they listened to the radio the entire hour-plus ride to his colleague's house.

He described the experience as, "we were like, yeah, I love Pink Floyd! We then put a CD in the player, cranked it up loud, put a hand on the dashboard, and rocked out the entire trip back to his place." This experience enunciates a point central to the field of sensory studies (e.g., Abram, 1997; Howes, 1991, 2005; Stoller, 1989, 1999)—in spite of a comparatively limited Western framing of the sensorium, replete with strong divisions between the senses, no sense truly operates in a solitary fashion and sensory scholarship is often stronger when recognizing more than one sense at a time. It also reaffirms the notion between sonority, sounds, and sociocultural constructs such as race (for example, Weheliye, 2004).

Thus, two central points emerge from this exploration of sound in curriculum studies. First, there is a history of significant, strong scholarship that can be placed under the umbrella of sound curriculum (e.g., Dimitriadis, 2009; Land & Stovall, 2009; Wang, 2007). However, the majority, if not all, of sound curriculum has utilized the visual form of text, organized sounds are rarely recognized as sounds but instead as talk and music, and there are relatively few examinations of sound-as-curriculum/as-pedagogy. This gives rise to the second central point of emergent understanding: There remains considerable work to be done in theorizing, utilizing, and examining sound curriculum. How, then, might sound curriculum contribute to contemporary studies of curriculum?

Silence, Noise, Sound Spaces, and Sound Art: The Field of Sound Curriculum

This section provides a few possible directions to the question posed above. To do so, I have placed ideas in sound curriculum in complicating conversations with scholarship drawn from the burgeoning field of sound studies (e.g., Drobnick, 2004; Brandt, Duffy, & MacKinnon, 2009; Erlmann, 2004). I begin with Miller (2005) and Jardine's (2004/1992) discussions of silence. This is followed in turn by a questioning of the inherently negative nature of noise, a complexifying of Aoki's (Pinar & Irwin, 2005) construction of sound spaces, and a very brief mention of the possibilities that sound art might bring to the field.

Silence and Noise: Social Control or Social Meaning?

As noted at the end of the previous section, silence in curriculum studies is conceived of as a positive place of reflection, natural quietude, or the often implicit quashing of a person or group's literal or metaphoric voice. In both instances, noise is silence's opposite. It is either the interrupting urban-ness of our daily lives that keeps us from finding our quiet places of solitude or the overwhelming sounds that disallow the space for a particular voice or voices to be heard.

In curriculum studies, examinations of silence and noise are often studies of social control. On the one hand, silencing is quite often a means of reifying dominant norms and values that seek to control local actors in educational contexts both intellectually and bodily, a tendency that is particularly the case for students and teachers in schools (e.g., Ingersoll, 2003; McNeil, 2000; Nespor, 1997). On the other hand, this silencing can be seen as a suppression of noise, the unwanted sounds that are understood to be either mechanical interruptions or students' interactions that lie outside of teacher-established, often narrowly defined, classroom participation structures (e.g., Erickson, 2004; Gutierrez, Rymes, & Larson, 1995).

Social studies of sound reaffirm this silence/noise dichotomy (Smith, 2004; Sterne, 2005). They do so, however, through explicit examinations of sound. Sterne (2005) for example, documents how Muzak is no longer simply "easy-listening 'background' sounds," but is now used as a "nonaggressive music deterrent" for a classed and raced consumerism (p. 7), a point to which I will return in the next section on sound spaces.

Yet, sound studies also contest these commonplace understandings of silence and noise in important ways. Cage (1973), whose notions of silence and sound still resonate with aspects of contemporary sound studies, is one such example.

> In fact, try as we may to make silence, we cannot. For certain engineering purposes, it is desirable to have as silent a situation as possible. Such a room is called an anechoic chamber. . . . I entered one at Harvard University several years ago and heard two sounds, one high and one low. When I described them to the engineer in charge, he informed me that the high one was my nervous system, the low one my blood in circulation. Until I die there will be sounds. And they will continue after my death.
>
> (p. 8)

Approaching a similar point from the opposite direction, in his discussion of the importance, possibilities, and power of sound as noise Attali (1985) writes, "Our science has always desired to monitor, measure, abstract, and castrate meaning, forgetting that life is full of noise and death alone is silent. . . . Nothing essential happens in the absence of noise" (p. 3).

In these and other such ways, silence and noise are social signals whose interpretation reveals local and less local norms and values. This point is well summarized by Bart Kosko (2007) in the introduction of his book *Noise*, "This book looks at noise both as something bad and something good. . . . The common goal in this noise inquiry is to see more noise patterns as signals whether or not we like those signals" (p. xviii). As with any other sensory information, both silence and noise contain important information as much about the listener's construction of that sound as they do the nature of the sound that is heard. With a few notable exceptions such as profoundly deaf individuals or artificial vacuums, human beings cannot turn off their ears and there is never any true silence.

Where a hearing person cannot help but hear, a physiological inevitability, the meanings one ascribes to those sounds are sociocultural constructions—nested layers of local and less local norms and values, combinations of particularized experiences, understandings, and tastes (Drobnick, 2004; Erlmann, 2004; Smith, 2004). It is important to note that sound is more than a social construction; sounds are an integral aspect of being human for the deaf and hearing alike. Although this chapter focuses on the meanings ascribed to sounds, overlooking the physiological for the social can lead to both misunderstandings about the nature of human conceptualizations of knowledges (Erlmann, 2010) and the ways in which deaf people are marginalized (Lane, 1999). Visual metaphors are certainly evocative, but they do not resonate in the same fashion.

Sound Spaces

While Aoki (Pinar & Irwin, 2005) and Jardine's (2004/1992) works present iterations of ecological space, Miller (2005) and others' works address questions of sound vis-à-vis silencing. Despite such difference in the locus of their discussions on sound, all also concern questions of physical and metaphoric space. Although such studies do examine both sound and space, they do not explore the ways in which sounds occupy or create physical, ecological spaces. Instead, curriculum studies of sound tend to denote the interrelatedness of sound (Jardine; Pinar and Irwin) or outline the metaphoric yet no less salient spaces in which one voice is silenced in favor of another or altogether ignored (e.g., Delpit, 1995; Weis & Fine, 2005; Miller; Thornton, 2004).

Similar in kind, yet rather different in focus, sound studies articulate the ways in which sounds literally create spaces (e.g., Blessing & Salter, 2009; Brand, Duffy, & MacKinnon, 2007; Born, 2015). This differs from how space is most often theorized in curriculum studies in general, an orientation that mirrors those who address sound/space outlined above, as either a physical (buildings, walls) or virtual/metaphorical phenomenon (boundaries between nation-states, people, or ideas) (e.g., Callejo-Perez, Slater, & Fain, 2003; Helfenbein & Taylor, 2009; Whitlock, 2007).

For example, Michael Bull (2001) notes how sounds create the physical boundaries of a car. Without audio playback via radio or compact disc (or, now, a variety of media for sound), people inside cars in motion are more likely to hear the environment in which they travel—tires on the road, cars passing, rain hitting the windshield. However, when an audio device is turned on, sounds from the speakers bounce off the physical boundaries of the car's interior, creating a much more contained space than the sonically porous walls and doors of a car usually provide. Additionally, sound in cars is not only aural but also physical, impacting more and more of the human body as volume is increased. In other words, although the physical space of cars is not altered, the literal space of cars, its boundaries and meanings, is profoundly transformed through sound.

Sterne's (2005) discussion of Muzak mentioned in passing above similarly enunciates how sounds can create boundaries that extend beyond the physical. In this instance, it is achieved by piping specific kinds of Muzak designed to simultaneously attract particular populations, while driving others away. Groups such as teens and the homeless in Sterne's piece, categories of unwanted people who were thought to both "loiter" and not fit the demographic of consumer that these stores desired, were specifically targeted for this kind of sonic assault, a point that fits in Goodman's (2010) larger argument about "sonic warfare."

Using Muzak in this manner extends the boundaries of a store beyond its physical boundaries to the limits of the speakers' volume as it flows out across parking lots and, in some instances, parks and other spaces adjacent to stores. In so doing, Muzak not only serves as a sonic marker of space, bounding one environment from another, it also imbues that space with a particular identity, a particular kind of store that values a given kind of consumer over another or a perceived non-consumer.

Qi et al.'s (2008) use of soundscapes to document both the boundaries and health of ecosystems is another example of how sound can be used to demarcate spaces and places. As the co-authors argue:

> Although . . . landscape characteristics, often derived from analysis of remotely sensed imager, are important indicators of ecosystem health, they are temporally static and do not provide a sufficient spatial resolution to observe the responses of individual organisms to anthropogenic disturbances. The audio characteristics emitted from an ecosystem, such as the sounds from birds, mechanical movements, or wind . . . provide unique insight into spatial and temporal patterns of ecosystem responses to human disturbances.
>
> (p. 201)

In sum, sounds create spaces that can further limit or supersede physical boundaries, a point addressed in the following chapter and brought to bear in previous work on sonic cartographies. They can also be used to denote historical time and value (e.g., Smith, 2004). Sounds similarly serve as markers between spaces, delineating one space or idea from another not unlike nightclubs that play themed music aimed to attract different groups in different rooms on the same night, or create place as further detailed in the following chapter.

Sound Art and Recording Culture

Although not as prevalent as visual art, sound art, most often in installations in public and gallery spaces, continues to grow. As there are already some excellent books attending to questions of sound/art (e.g., Kahn, 2001; Kim-Cohen, 2009; LaBelle, 2015; Licht & O'Rourke, 2007; Voegelin, 2010;

Weibel, 2016), it is important to note how intentionally organized sound spaces can represent particular ideas and ideals in non-visual or not primarily visual ways (e.g., Kahn, 2001). Additionally, there is growing attention to the relationships between sound and social science methodology, particularly between audio recording and ethnography (Drever, 2002; Samuels, Meintjes, Ochoa, & Porcello, 2010; Chapter 5). As this work indicates, particularly in light of having reached a point in the distribution of technology where most if not all personal computers have digital audio playback programs, sound curriculum, and empirical sonic studies, can now be disseminated sonically with as much ease as textual representations.

Implications and Possibilities: Moving Beyond the Frame

What are the implications for the future of sound curriculum? How might they contribute to the field? In short, so what? To answer these questions requires starting once again from the top in order to trace these successive iterations of sound curriculum as it is framed and reframed, to move from what was always already there to a sketch of possible future sound directions for the field that lie beyond the frame.

As outlined in the first section, a nascent sound curriculum already exists in contemporary curriculum studies. This work includes scholarship that addresses the importance of talk, music, and sound as forms of knowledge. Although most of this scholarship tends to regard talk and music not as sound but as discreet expressions and ways of knowing, it does suggest that further, more focused, work in these directions might be undertaken. For example, lyrics to music and music without lyrics might be more explicitly studied and used pedagogically, particularly in light of the fact that musicians are indeed public intellectuals who pass knowledge onto others (Gershon, 2010). It also suggests that talk and music might be theorized and otherwise examined as sound, a move that could serve to make the familiar strange and, in so doing, help curricularists better listen to students, teachers, and educational contexts.

Complicating this conversation are notions that silence is not without sound, noise is not without pattern or meaning, and sounds create spaces within and past physical boundaries. From this perspective, school bells not only time the bodies in schools but mark time for the city blocks that surround them, as the sound of the bells bounces off concrete, over fields, and through windows. This also has implications for questions of silence and silencing that could be used to further demonstrate one's agency even in subjugated, marginalized positions. Similarly, in a time when the question often seems to be how to better reach those outside of the scholarly community, the use of sound installations and public sound art seems to be a rather viable solution.

Yet, these constructions ultimately remain bounded, held fast by visual ways of knowing and assumptions about the applicability of ocular metaphors to non-sighted understandings. At an intellectual moment when differences between unintentional hearing and purposeful listening are being challenged, the idea of *framing* sonic constructions of meaning seems increasingly inappropriate. How, then, might sounds be conceptualized?

One possible answer to this question resides in an examination of discourse in curriculum studies. As scholarship in sociolinguistics has demonstrated, metaphors both give voice to our conceptualizations of our worlds and form frameworks of understanding (e.g., Blom & Gumperz, 1972; Duranti, 2001; Lakoff & Johnson, 2003). It is therefore of utmost importance that discourse evolves with an expanding notion of sonic knowledge and aural understandings. Scholarship presented throughout this chapter provides some possible paths in this direction. For example, Erlmann's (2010) strong analysis of a missing otological throughline (den Heyer, 2005) in Western philosophy underscores the power and possibility for considerations of resonance, a point taken up here in Chapter 3. Similarly, Sterne (2005), Feld (2005), and other examinations at the intersection of sound and space suggest that the kinds of language used by fields such as critical geography might also be a good fit for discussions of sound. In addition to such possibilities as Feld's talk of acoustemology and terminology associated with the construction of soundscape (Samuels, Meintjes, Ochoa, & Porcello, 2010), consider, for example, what it might mean to begin to examine *sound fields*. Along similar lines, such a turn might be to take seriously aural terminology that often finds its way into scholarly discourse (i.e., pitch, tone, tenor, polyphony) as literal examinations of sound rather than as metaphorical markers of differentiation. Furthermore, one could ask what it might mean to listen closely to an environment rather than give it a close reading: How does sound advice resonate?

A second possibility lies in the physicality of sound, its embodied nature. Sounds occupy a unique place in human understanding in their dichotomous presence, sounded outside and perceived inside the body. So present is our perception of sound that its embodiment is a touchstone not only of our physical selves but also of the ways in which we have constructed our virtual and recorded selves, in film and digital media for example. As such, sounds occupy a space that is beyond constructions of the postmodern. It both resonates inside us and resounds outside our persons, knowledge that we construct through everyday experience that is neither unnameable nor named. From this perspective, sound is not only something heard but also something felt, a phenomenon that places aurality and sonic meanings at the forefront of questions in the field of curriculum studies.

Finally, it is precisely these emergent sonic meanings that may provide some of the more groundbreaking possibilities in curriculum studies. It may well be the case that sounds are best represented sonically, beyond the text and, perhaps, without visual accompaniment. What might it mean for ideas to be represented as sounds? How might a person's agency be altered when a reader becomes a listener and can hear that person speak in her own voice? What would it mean for a listener to hear the sounds of classrooms, of knowledge being constructed, that the multitude of noises that combine to create a given sonic environment?

It is within these possibilities that a sonic curriculum strikes new ground beyond current frames and frameworks in curriculum studies. Not only can a sound curriculum assist in making familiar discourses, contexts, and interactions strange, such aurality could also embody the ways in which the unfamiliar might resonate across what initially seem to be disparate or irreconcilable differences. A sound curriculum similarly holds the possibility of a set of relatively unexplored discourses and practices that address questions that reside in our current curriculum quandary that I have come to think of as "post-next," the difficulties in describing our contemporary moment that moves beyond post- constructions of understanding (see, for example, www.naparconference.asu.edu/).

Sound meanings also question the ocular nature of frames and their ability to respond to other sensual ways of knowing, in this case, embodied sonic knowledge. In these ways, attention to sound curriculum simultaneously reifies the possibilities of framing ideas and ideals in and as sound while interrupting taken-for-granted notions about how frames delineate this from that, one thing from another. For example, where Derrida (1998) famously locates hearing and the tympan (eardrum) as the source of splitting self from other, Erlmann (2010) questions both the biological and metaphorical sound-ness of Derrida's argument—the eardrum does not reflect and separate, instead, it receives and resonates. This is not to say that frames are not useful or that Butler's (2009) extension of frames as empowered by their iterative removal and reconstitution from their original context and notions of othering through framing is somehow invalidated or otherwise unhelpful. However, these and other previous constructions of framing (e.g., Bateson, 1972) nonetheless tend to overlook the possibilities non-visual (and non-Western) metaphors and epistemologies can provide, remaining grounded in the very kinds of static constructions Butler argues against.

In sum, the importance of a sound curriculum lies in the opportunities it provides for conceptualizing curricular relationships, interactions, and meanings. Such opportunities retain the critical, multifaceted aspects of current curricular understandings. They also add key dimensions to conceptual, methodological, and practical frameworks that create the space for now-familiar ways of understanding curriculum to be rendered strange in order that they might be better heard, critiqued, and translated, themes that echo throughout this book. By opening ourselves to the possibilities of sound curriculum—explorations of the sonic in education as well as the uses and possibilities provided by sound/education—we can not only maintain our current critical and creative endeavours but also approach them from underutilized ways that will let us truly hear what educative contexts have to say, provided that we listen.

Organization and Contents

The remainder of this book has been organized in the following manner. Chapter 2 continues with themes about sound education raised in this introductory chapter, particularly those that attend to considerations of both the ways in which sounds form educational systems of knowledge and the construction of educational places and spaces. Chapter 3 addresses how a philosophy of educational resonances might simultaneously interrupt commonly held educational understandings while providing a richly enmeshed onto-epistemogenic platform for the consideration and understanding of educational interactions.

Where Chapter 4 attends to the idea that we wish the hearing deaf and the deaf hearing. Chapter 5 returns to a theme I first explored nearly a decade ago, that of the lens of improvisation as a means to conceptualize educational interactions. Where my previous work on the topic sought to open improvising metaphors for wondering about moments of classroom interactions, here I argue that all students, and by extension their teachers, are improvisers and all educational interactions are forms of improvisation.

Chapter 6 speaks across previous publications on how sound might function in qualitative research, a consideration of the kinds of ideas, ideals, and practices that are often central to conducting research in, as, or through sound. Where the previous chapters use qualitative studies of sound to explicate theoretical, philosophical, and methodological arguments, the last two chapters before the brief chapter that concludes this book are both empirical in nature. Chapter 7 tells a tale often avoided in discussions of how music can be used as a curricular tool, an inversion of the possibilities of songs to understand academic content. Here, a rich, critically creative and collaborative process becomes relatively innocuous and impotent knowledge. Chapter 8 in turn flips this on its head, documenting how sound social science can become powerful forms of sound/art that fill spaces with overlooked narratives and, in the process, are iterations of social justice that reverberate beyond the boundaries in which they are played.

These two final chapters also serve as examples for two of the many ways that sonic scholarship can function. Where Chapter 7 on music as a curricular tool is primarily texted and sounds function in support of the text, Chapter 8 is primarily sonic and the text serves to further explicate the text. In spite of these differences, both chapters also share a sonic construction in that they each ends with a sonic expression of the chapter. A final brief concluding chapter returns to themes presented throughout the book in order to argue for the potential significance of Pauline Oliveros' (2005) work in sound curriculum, listening deeply to children and youth as a central educational practice.

The sounds found throughout this book can be heard by either selecting the hyperlinked text in the e-versions of this work or copying the associated web address (URL) into a browser of your choice. Although many of the points can certainly be ascertained by simply reading the text, they are nonetheless lacking and incomplete without attending to the associated sound files. A brief note about the quality of the sound files found throughout this book: Mp3 files compress sounds in much the way that one might squish a piece of bread, and with similar results—highs and lows are pressed closer together and the middle becomes more mushy. However, given the ubiquity of mp3 playback and the increase in speed and smoothness their compression brings listeners, not to mention how so many of us have gotten used to this compression as "normal," I have elected to use mp3 files here so that access trumped quality (for what is the use of having sound files that fewer people can actually hear?).

Finally, this book cuts across what are often different scholars' foci in curriculum studies, sound studies, and in social sciences and humanities more generally. By this I mean that the contents work from theoretical to methodological to empirical in messy, nonlinear ways where parts of empirical studies explicate theoretical positions and philosophical concerns are all wrapped up with empirical evidence. As noted in the preface, although the chapters have indeed been intentionally aligned to move from theory to philosophy to method to empirical, each of these chapters is also in many ways presented unto itself. It is therefore at the reader's discretion on how to approach this book, as is always the case. In other words, don't let the organization of the book get in the way of how you read it or the words get in the way of your ears.

References

Abram, D. (1997). *The spell of the sensuous: Perception and language in a more-than-human world*. New York: Vintage.

Agar, M. H. (1996). *The professional stranger: An informal introduction to ethnography* (2nd ed.). San Diego, CA: Academic Press.

Agar, M. H. (2004). We have met the other and we're all nonlinear: Ethnography as a nonlinear dynamic system. *Complexity*, *10*(2), 16–24.

Appelbaum, P. (2007). *Children's books for grown-up teachers: Reading and writing curriculum theory*. New York: Routledge.

Aoki, T. T. (1991). Sonare and vidare: Questioning the primacy of the eye in curriculum talk. In G. Willis & W. H. Schubert (Eds.), *Reflections from the heart of educational inquiry: Understanding curriculum and teaching through the arts*. (pp. 82–89). Albany, NY: State University of New York Press.

Attali, J. (1985). *Noise: The political economy of music* (B. Massumi, trans.). Minneapolis, MN: University of Minnesota Press.

Au, W., Brown, A. L., & Calderón, D. (2015). *Reclaiming the multicultural roots of U.S. curriculum: Communities of color and official knowledge in education*. New York: Teachers College Press.

Back, L. (2007). *The art of listening*. Oxford: Berg Press.

Bateson, G. (1972). *Steps to an ecology of mind*. Chicago: University of Chicago Press.

Blessing, B., & Salter, L.-R. (2009). *Spaces speak, are you listening?: Experiencing aural architecture*. Durham, NC: Duke University Press.

Blom, J., & Gumperz, J. (1972). Social meaning in linguistic structure: Code switching in Norway (pp. 407–434). In J. Gumperz & D. Hymes (Eds.), *Directions in sociolinguistics: The ethnography of communication*. New York: Holt, Reinhart & Winston.

Borges, J. L. (1998). The garden of forking paths. In A. Hurley (Trans.), *Collected fictions* (pp. 119–128). London: Allen Lane.

Born, G. (2015). *Music, sound and space: Transformations of public and private experience*. Cambridge, UK: Cambridge University Press.

Brandt, R., Duffy, M., & MacKinnon, D. (Eds.). (2009). *Hearing places: Sound, place, time and culture*. Cambridge, UK: Cambridge Scholars Press.

Bull, M. (2001). Soundscapes of the car: A critical study of automobile habitation. In D. Miller (Ed.), *Car cultures* (pp. 185–202). Oxford: Berg Press.

Butler, J. (2009). *Frames of war*.L ondon: Verso.

Cage, J. (1973). *Silence: Lectures and writings*. Middletown, CT. Wesleyan University Press.

Callejo-Perez, D. M., Fain, S. M., & Slater, J. J. (Eds.). (2003). *Pedagogy of place: Seeing space as cultural education*. New York: Peter Lang.

Castenell, L. A. Jr., & Pinar, W. F. (Eds.). (1993). *Understanding curriculum as racial text: Representations of identity and difference in education*. Albany, NY: SUNY Press.

Cooper, A. C. (1892). *A voice for the South by a Black woman from the South*. Xenia, OH: Aldine Printing House.

Dell'Antonio, A. (Ed.). (2004). *Beyond structural listening? Postmodern modes of hearing*. Berkeley, CA: University of California Press.

Delpit, L. (1995). *Other people's children: Cultural conflict in the classroom*. New York: New Press.

Derrida, J. (1998). *Of grammatology* (G. C. Spivak, trans.). Baltimore, MD: Johns Hopkins University Press.

Dimitriadis, G. (2009). *Performing identity/ performing culture: Hip hop as text, pedagogy, and lived practice* (Revised ed.). New York: Peter Lang.

Dockrell, J. E., & Shield, B. M. (2006). Acoustical barriers in classrooms: The impact of noise on performance in the classroom. *British Educational Research Journal, 32*(3), 509–525.

Drever, J. L. (2002). Soundscape composition: The convergence of ethnography and acoustamatic music. *Organised Sound*, 7(1), 21–27.

Drobnick, J. (Ed.). (2004). *Aural cultures*. Toronto, ON: YYZ Books.

DuBois, W. E. B. (1903). *The souls of Black folk*. Chicago: McClurgand.

DuBois, W. E. B. (1926). *Criteria of negro art*. Retrieved from: www.webdubois.org/db CriteriaNArt.html

Duranti, A. (Ed.). (2001). *Linguistic a nthropology: A reader*. Malden, MA: Wiley-Blackwell.

Eisner, E. W. (1985). *The educational imagination: On the design and evaluation of school program* (2nd ed.). New York: Palgrave Macmillan.

Eisner, E. W. (2002). *The arts and a creation of mind*. New Haven, CT: Yale University Press.

Ellsworth, E. A. (2005). *Places of learning: Media, architecture, pedagogy*. New York: Routledge Falmer.

Erickson, F. (1982). Classroom discourse as improvisation: Relationships between academic task structure and social participation structure in lessons. In L. C. Wilkinson (Ed.), *Communicating in the classroom* (pp. 153–181). New York: Academic Press.

Erickson, F. (2003). Some notes on musicality and speech. In D. Tannen & J. E. Atalis (Eds.), *Linguistics, language and the real world: Discourse and beyond (Georgetown University:*

Round table on languages and linguistics) (pp. 11–35). Washington, DC: Georgetown University Press.

Erickson, F. (2004). *Talk and social theory: Ecologies of speaking and learning in everyday life.* Malden, MA: Polity Press.

Erlmann, V. (Ed.). (2004). *Hearing cultures: Essays on sound, listening and modernity.* Oxford: Berg Press.

Erlmann, V. (2010). *Reason and resonance: A history of modern aurality.* Cambridge, MA: Zone Books.

Feld, S. (1996). Waterfalls of song: An acoustemology of place resounding in Bosavi, Papua New Guinea. In S. Feld & K. H. Basso (Eds.), *Senses of place* (pp. 91–135). Santa Fe, NM: School of American Research Press.

Feld, S. (2005). Places sensed, senses placed: Toward a sensuous epistemology of environments. In D. Howes (Ed.), *Empire of the senses: The sensual culture reader* (pp. 179–191). Oxford: Berg Press.

Feld, S., & Basso, K. H. (Eds.). (1996). *Senses of place.* Santa Fe, NM: School of American Research Press.

Gaztambide-Fernández, R. (2010). Wherefore the musicians? *Philosophy of Music Education Review, 18*(1), 65–84.

Gaztambide-Fernández, R. (2011). Musicking in the city: Reconceptualizing urban music education as cultural practice. Special Issue: Music education in urban contexts. *Action, Criticism, and Theory for Music Education. 10*(1): 15–46.

Gershon, W. S. (2010). Entertaining ideas and embodied knowledge: Musicians as public intellectuals. In J. A. Sandlin, B. D. Schultz, & J. Burdick (Eds.), *The handbook of public pedagogy* (pp. 628–638). New York: Routledge.

Gershon, W. S. (2011, January 23). Sonic ethnography: Sound meanings,meaningful sounds. Paper presented at the Narrative, Arts-Based, & "Post" Approaches to Social Research Conference (NAPAR), Tempe, AZ.

Gershon, W. S. (2016). The sound of silence: The material consequences of scholarship. In N. Snaza, S. E. Truman, D. Sonu, & Z. Zowiska (Eds.), *Pedagogical matters: New materialisms and curriculum studies.*(pp. 75–90). New York: Peter Lang.

Gershon, W. S. (In Press). *Sensuous curriculum: Politics of the senses in education.* Charlotte, NC: Information Age Press.

Goodman, S. (2010). *Sonic warfare: Sound, affect, and the ecology of fear.* Cambridge, MA: MIT Press.

Gouzouasis, P. (2008). Toccata on assessment, validity and interpretation. In S. Springgay, R. L. Irwin, C. Leggo, & P. Gouzouasis (Eds.), *Being with a/r/tography* (pp. 221–232). Rotterdam, NL: Sense Publishers.

Grant, G. A., Brown, K. D., & Brown, A. L. (2015). *Black intellectual thought in education: The missing traditions of Anna Julia Cooper, Carter G. Woodson, and Alain LeRoy Locke.* New York: Routledge.

Greene, M. (1995). *Releasing the imagination: Essays on education, art, and social change.* San Francisco: Jossey-Bass.

Gustafson, R. I. (2008). Drifters and the dancing mad: The public school music curriculum and the fabrication of boundaries for participation. *Curriculum Inquiry, 38*(3), 267–297.

Gustafson, R. I. (2009). *Race and curriculum: Music in childhood education.* New York: Palgrave Macmillan.

Gutierrez, K., Rymes, B., & Larson, J., (1995). Script, counterscript, and underlife in the classroom: James Brown versus Brown v. Board of Education. *Harvard Educational Review, 65*(3), 445–471.

Handelman, D. (2001). Framing, braiding, and killing play. *Focaal, 37*, 145–158.

Helfenbein, R. J., & Taylor, L. H. (2009). Critical geographies in/of education. [Special edition] *Educational Studies, 45*.

Heyer, K. den (2005). To what questions are schools answers? And what of our courses? Animating throughline questions to promote students' questabilities. *Canadian Social Studies, 39*(2). Retrieved from http://www2.education.ualberta.ca/css/css-39_2/ARden Heyer_throughline_questions.htm

Houseman, M. (2001). Is this play? Hazing in French preparatory schools. *Focaal: The European Journal of Anthropology, 37*, 39–48.

Howes, D. (Ed.). (1991). *The varieties of sensory experience: A sourcebook in the anthropology of the senses*. Toronto: University of Toronto Press.

Howes, D. (2003). *Sensual relations: Engaging the senses in culture & social theory*. Ann Arbor, MI: University of Michigan Press.

Howes, D. (Ed.). (2005). *Empire of the senses: The sensual culture reader* (pp. 179–191). Oxford: Berg Press.

Ihde, D. (2007). *Listening and voice: Phenomenologies of sound* (2nd ed.). Albany, NY: State University of New York Press.

Ingersoll, R. N. (2003). *Who controls teachers' work? Power and accountability in America's schools*. Cambridge, MA: Harvard University Press.

Jardine, D. (2004/1992). A bell ringing in the empty sky. In W. F. Pinar (Ed.), *Contemporary curriculum discourses: Twenty years of JCT* (pp. 262–277). New York: Peter Lang.

Kahn, D. (2001). *Noise water meat: A history of sound in the arts*. Cambridge, MA: MIT Press.

Kosko, B. (2007). *Noise*. N ew York: Viking.

LaBelle, B. (2010). *Acoustic territories: Sound culture and everyday life*. New York: Continuum.

LaBelle, B. (2015). *Background noise: Perspectives on sound art* (2nd ed.). New York: Bloomsbury.

Lakoff, G., & Johnson, M. (2003). *Metaphors we live by* (2nd ed.). Chicago: University of Chicago Press.

Land, R. R., & Stovall, D. O. (Eds.). (2009). Hip hop and social justice. [Special Edition]. *Equity and Excellence in Education, 42*(1).

Licht, A., & O'Rourke, J. (2007). *Sound art: Beyond music, between categories*. New York: Rizzoli.

Lindquist, G. (2001). Elusive play and its relations to power. *Focaal: European Journal of Anthropology, 37*, 13–23.

McCarthy, C., Hudak, G., Miklaucic, S., & Saukko, P. (Eds.). (1999). *Sound identities: Popular music and the cultural politics of education*. New York: Peter Lang.

McNeil, L. (2000). *Contradictions of school reform: Educational costs of standardized testing*. New York: R outledge.

Malewski, E. (2009). *Curriculum studies handbook: The next moment*. New York: Routledge.

Miller, J. (2005). *Sounds of silence breaking: Women, autobiography, curriculum*. New York: Peter Lang.

Morris, M. (1999). Curriculum as musical text. In M. Morris, M. A. Doll & W. F. Pinar (Eds.), *How we work* (pp. 11–18). New York: Peter Lang.

Nespor, J. (1997). *Tangled up in school: Politics, space, bodies and signs in the educational process*. Mahwah, NJ: Lawrence Erlbaum.

O'Callaghan, C. (2007). *Sounds: A philosophical theory*. Oxford, UK: Oxford University Press.

Oliveros, P. (2013). *Anthology of text scores*. Bloomington, IN: iUniverse: Deep Listening Publications.

Pendergast, M., Leggo, C., & Sameshima, P. (Eds.). (2009). *Poetic inquiry: Vibrant voices in the social sciences*. Rotterdam, NL: Sense Publishing.

Piaget, J. (1962). *Play, dreams, and imitation in childhood.* New York: Ballantine.

Pinar, W. F., & Irwin, R. L. (2005). *Curriculum in a new key: The collected works of Ted Aoki.* Mahwah, NJ: Lawrence Erlbaum.

Pinar, W. F., Reynolds, R., Slattery, P., & Taubman, P. (Eds.). (1995). *Understanding curriculum: An introduction to the study of historical and contemporary curriculum discourses.* New York: Peter Lang.

Qi, J., Gage, S. H., Joo, W., Napoletano, B., & Biswas S. (2008). Soundscape characteristics of an environment: A new ecological indicator of ecosystem health. In W. Ji (Ed.), *Wetland and water resource modeling and assessment: A watershed perspective* (pp. 201–211). Boca Raton, FL: CRC Press/Taylor & Francis Group.

Rosario, J. (1991). On thinking as a sacred act, Coltrane jazz, the inaccessible, and curriculum. In G. Willis & W. H. Schubert (Eds.), *Reflections from the heart of educational inquiry: Understanding curriculum and teaching through the arts* (pp. 174–181). Albany, NY: SUNY Press.

Samuels, D. W., Meintjes, L., Ochoa, A. M., & Porcello, T. (2010). Soundscapes: Toward a sounded anthropology. *Annual Review of Anthropology, 39*, 329–345.

Sauer, T. (2009). *Notations 21.* New York: Mark Batty Publisher.

Serra, M. R., & Biasooni, E. C. (1998). Urban noise and classroom acoustical conditions in the teaching and learning process. *Journal of Environmental Studies, 56*, 41–59.

Shield, B. M., & Dockrell, J. E. (2003). The effects of noise on children at school: A review. *Building Acoustics, 10*(2), 97–116.

Smith, M. M. (Ed.). (2004). *Hearing history: A reader.* Athens, GA: University of Georgia Press.

Spindler, G., & Hammond, L. (2006). *Innovations in educational ethnography: Theory, methods and results.* Mahwah, NJ: Lawrence Erlbaum.

Springgay, S. (2002). Arts-based educational research as an unknowable text. *Alberta Journal of Educational Research, 48*(3). 1–30.

Springgay, S., Irwin, R., Leggo, C., & Gouzouasis, P. (Eds.). (2007). *Being with a/r/tography.* Rotterdam: Sense Publishers.

Sterne, J. (2005). Urban media and the politics of sound space. *Open/Sound, 9*, 6–14.

Sterne, J. (2012). *MP3: The meaning of a format.* Durham, NC: Duke University Press.

Stoller, P. (1989). *The taste of ethnographic things: The senses in anthropology.* Philadelphia: University of Pennsylvania Press.

Stoller, P. (1997). *Sensuous scholarship.* Philadelphia: University of Pennsylvania Press.

Tarc, A. M. (2015). *Literacy of the Other: Renarrating humanity.* Albany, NY: SUNY Press.

Thornton, S. J. (2004). Silence on gays and lesbians in the social studies curriculum. In S. J. F. Thornton and D. J. Flinders (Eds.), *The curriculum studies reader* (2nd ed., pp. 307–314). New York: Routledge.

Valli, L., & Chambliss, M. (2007). Creating classroom cultures: One teacher, two lessons, and a high-stakes test. *Anthropology Education Quarterly, 38*(1), 57–75.

Voegelin, S. (2010). *Listening to noise and silence: Towards a philosophy of sound art.* New York: Continuum.

Vygotsky, L. S. (1976). Play and its role in the mental development of the child. In J. S. Bruner, A. Jolly, & K. Silva (Eds.). *Play: Its role in development evolution* (pp. 537–544). New York: Basic Books.

Wang, H. (2007). The strength of the feminine, lyrics of Chinese women's subjectivity, and the power of education. In C. Eppert & H. Wang (Eds.), *Cross-cultural studies in education: Eastern thought, educational insights* (pp. 313–333). Mahwah, NJ: Lawrence Erlbaum.

Watkins, W. H. (2001). *The white architects of Black education: Ideology and power in America, 1865–1954*. New York: Teachers College Press.
Weibel, P. (Ed.). (2016). *Soundart: Sound as a medium of art*. Cambridge, MA: MIT Press.
Weis, L., & Fine, M. (2005). *Beyond silenced voices: Class, race, and gender in United States schools*. (3rd ed.). Albany, NY: State University of New York.
Weheliye, A. G. (2004). *Phonographies: Grooves in sonic Afro-modernity*. Durham, NC: Duke University Press.
Whitlock, R. U. (2007). *This corner of Canaan: Curriculum studies of place & the reconstruction of the South*. New York: Peter Lang.
Wozolek, B. W. (2015). *The presence of absence: The negotiation of space and place for students of color at a predominantly white suburban high school*. Dissertation. Kent State University.

2 Sounds as Educational Systems of Meaning

No matter where you are or what you're doing, you're gaining sonic information. And, regardless of what those sounds mean or their impact on your person, just as you are but a small part of a deep matrix of sonority, you are at once a node in sound ecologies and a contributor to these ever-emergent, undulating ambient vibrations, always a medium and media, a point to which I return in greater depth in the next chapter. Here, the focus is on the argument that sounds, when they hit living things from trees to babies, are, inescapably and always, forms of embodied knowledge.[1]

Where a hearing person cannot help but hear, a physiological inevitability, the meanings one ascribes to those sounds are sociocultural constructions—nested layers of local and less local norms and values, combinations of particularized experiences, understandings, and tastes (e.g., Erlmann, 2004; Gandy & Nilsen, 2014; Smith, 2004).[2] As such, sounds combine to form systems of meaning that can serve to simultaneously transmit, reify, challenge, or reinvent sociocultural norms and values (Brandt, Duffy, & MacKinnon, 2009; Feld, 1991, 1996).

In these ways, sounds carry knowledge from source to listener (Aoki, 1991; Erickson, 2004). As I argue in this chapter, sounds are therefore necessarily educational in nature, sensuous data so rife with information that the listener can render often disparate-seeming sonic input into embodied meaning systems. In short, to paraphrase Geertz (1983), sounds are systems of educational ways of beingknowingdoing. Given the combination of the sociocultural nature of sound meanings and their ability (if not inevitability) to educate, sounds therefore serve as a way to both understand the knowledge a group values and the values embedded in those meanings.

While this notion may appear relatively straightforward at its surface, it is a constellation of interconnected complexity and nuance. For example, scholars note an important distinction between the relatively passive and largely inevitable act of hearing and the active, purposeful intention of listening (e.g., Back, 2007; Bull & Back, 2003a), conceptualization of sound that Erlmann (2010) challenges as maintaining structuralist binaries, even when troubled from a postmodern perspective (p. 22–23) and further examined by Anahid Kassabian (2015, especially pp. xi–xxviii). Similarly, although noise may be considered anathema to learning (e.g.,Dockrell & Shield, 2006; Shield & Dockrell, 2003), there has been little attention paid to the meanings in noises in the consideration that they might be educative (for discussions of noise check, for example, Attali, 1985; Voegelin, 2010).

However, as the burgeoning field of sound studies has demonstrated, sounds of all kinds are part and parcel of how people literally make sense[3] of their worlds. Sounds similarly inform the ways in which people understand their relation to the spaces and places, understandings that strongly contextualize the ways in which individuals constitute their identities (Brandt, Duffy, & MacKinnon, 2009; Bull, 2001; Feld, 1996; Gershon, 2013).

The aim of this chapter is therefore to demonstrate how sounds combine to create embodied educational systems of meaning regardless of their apparent organization or categorization according to local norms and values (e.g., talk vs. noise). While there has been some work in education that documents the musical nature of classroom interactions (e.g., Erickson, 1982, 2003; Gershon, 2006) and a rather long history of examining classroom talk (e.g., Bellack, Kliebard, Hyman, & Smith, 1978; Erickson, 2004; Heath, 1983), there has been scant study of *sound* in educational contexts either in or out of schools, other than negatively connoted as a distraction to learning or other significant educational ways of being.

In order to document how sounds form embodied educational meaning systems, I begin with an expansion of the points raised in this brief introduction, a discussion that includes an overview of both how sounds are most often regarded in education and the ways in which sounds are now often conceptualized in the burgeoning field of sound studies. This is followed by a short transcribed vignette of an urban fifth grade class in transition that documents how educational sounds are most often regarded in classrooms and the shortcomings of such interpretations. I then present an altogether different educational context—another group of fifth graders from the same classroom doing a sensory walk in the Cuyahoga Valley National Park—in order to note that arguments made about sound knowledge, while certainly informed by their contexts, largely hold fast across (within and between) sonic ecologies. The final concluding section suggests some implications in light of these sound ideas.

Sounds as Embodied Knowledge in Curriculum and Sound Studies

This section provides an overview of the literature on sounds as embodied knowledge in curriculum studies and the still-emergent, interdisciplinary field of sound studies. Although aspects of this argument were addressed in Chapter 1, I have elected to again attend to points that are central to such understandings due to differences in the focus of my argument, the relative newness of considering scholarship in the educational literature as sound curriculum and the general underrepresentation of sound studies in education.

In spite of concerns raised to the contrary and calls to listening and awareness as noted in the previous chapter (e.g., Appelbaum, 2007; Aoki, 1991; Jardine, 2004), the field of curriculum studies most often reads sounds as Texts, in no small part a result of both a prevailing tendency in the field and across the social sciences to render ideas, spaces, and interactions in this fashion (e.g., Gershon, 2013; McCartney, 2012; Shippins & Grant, in press; also Chapter 5). The advantage to conceptualizing knowledge as a Text is that it creates the discursive space to deconstruct and otherwise render non-texts as one would critically approach a written text, revealing explicit themes, hidden motifs, and other such explicit and implied patterns of meaning.

However, conceptualizing sound as text has two major implications for understanding sounds as embodied educational systems of meaning. First, the consideration of sounds as texts has provided a strong foundation for understanding formal, hidden, and enacted curriculum. Consider, for example, the ways in which early educational ethnographies were in many ways studies of language use, examinations that led scholars to theorize about the strangeness in familiar patterns of schooling (e.g., Heath, 1983; Mehan, 1979; Spindler, 1982) and the central role of language, talk, and text in postmodern and post-structural scholarship in the field (e.g., Peters, Marshall, & Fitssimons, 1999; Slattery, 2006). The second result of conceptualizing sounds as texts in curriculum studies has been a focus on sounds as texts of talk and music, sounds I refer to as organized for the ways in which they are intentionally assembled (e.g., Erickson, 2004; Landy, 2007).

Yet, while there have been discussions of music in curriculum studies (e.g., Appelbaum, 2007; Dimitriadis, 2009; Gustafson, 2009; McCarthy, Hudak, Miklaucic, & Saukko, 1999; Morris, 1999), they are far more infrequent than discussions of "fine art" (painting, sculpture, etc.), new media, and the implications of art for the field such as Eliot Eisner's (1977) now famous argument for educational assessment as artistic critique. Furthermore, discussions of sound, as opposed to its categorization as texted talk and music, are few and far between. My own writing that focused on embodied knowledge and music rather than on embodied sound knowledge is but one recent example (Gershon, 2011a).

Ted T. Aoki's life of scholarship (Pinar & Irwin, 2005) is one of the few exceptions to this tendency, as are Fredrick Erickson's (1982, 2003) work that examines the nexus between talk, music, and meaning; David Jardine's (2004) call to mindfulness; and Maxine Greene's (1995; Ayers & Miller, 1997) impact on the field of curriculum studies in general and her attention to the sensuous, visual, and sonic ways of knowing in specific. Yet, now nearly 20 years past his call to sonic awareness, Aoki's (1991) concern over "the primacy of the eye in curriculum studies" still has gone largely unheeded.

Therefore, although there have been instances of scholarship that have approached the possibility of sounds as embodied knowledge (Jardine, 2004; Pinar & Irwin, 2005), such scholarship is decidedly infrequent when compared to iterations of sounds-as-text in curriculum studies. Curriculum studies is by no means alone in its textualization of the sonic and/or of the senses in general, this in spite of a steep uptick in educational scholarship that attends to or is grounded in the work of Giles Deleuze and Felix Guattari, especially their work *A Thousand Plateaus* (1987) and its accompanying attention to what can be thought of as the affective turn (Clough, 2007) in the humanities and social sciences (e.g., Clough & Halley, 2007; Gregg & Seigworth, 2010; Stewart, 2007; Massumi, 2015). This focus on text can be understood as following the literary turn and subsequent moves to read interactions and ideas as texts in the social sciences, a move David Howes (2003) characterizes as moving "from reading texts to writing culture" (p. 22–26).

Contrary to such curricular understandings, the field of sound studies is characterized in no small part by disparate sonic trajectories of scholarship, pathways that are articulated in equally diverse ways with a commitment to open, fluid borders and sound interpretations. Then there are questions of field and its placement. Although sound studies can certainly be understood to share some of its borders with other sensory-related fields as well as the fields of affect theory and sensory studies, its inter/trans/disciplinary nature provides a seemingly endless array for sonic exploration. Nevertheless, my purpose here is to provide an overview of some of the broader trends within this still-emerging field, possibilities that provide powerful means for attending to and expressing experiences, events, relationships, people, things, and ecologies. Such possibilities are also important because they can and will continue to lead to pathways for reconceptualizing such phenomena as well as the reconsideration of the norms and values that contextualize them.

Before continuing, it is important to further situate this chapter within broader contemporary constructions of sound meanings and their relationships to other sensuous knowledge. As scholars invested in sound and sensory ways of knowing have underscored (e.g., Drobnick, 2004; Erlmann, 2004; Gershon, 2011b; Howes, 2003, 2005), sensuous understandings are often the most powerful when more than one sense is used in concert with another. Along similar lines, my argument here is not an effort to elevate sound over any other sensual understandings regardless of their categorization or construction. Instead, it is designed to follow Aoki in questioning the primacy of vision in curriculum studies by providing an illustration of the ways in which sounds form educational systems of embodied meanings.

Sound Meanings

Sound studies comprises "the disparate threads of an ever-expanding field of writing on the social nature and meaning of sound . . . from the fields of sociology, cultural studies, media studies, anthropology, cultural history, philosophy, urban geography and musicology" (Bull & Back, 2003b, p. 3)—an interdisciplinary and complex conversation that is not dissimilar to contemporary curriculum studies (e.g., Malewski, 2009; Pinar, 2004). To these fields I would also add those of ethnomusicology, sociolinguistics, organized sound, and sound art that are similarly involved in questions that regard the relationship between sound and meaning (e.g., Kahn, 2013; Kim-Cohen, 2009; LaBelle, 2010; Landy, 2007).

Yet, as Erlmann (2004) notes,

> [e]ven in ethnomusicology and musicology—two disciplines that might lay superior claim to sound and auditory perception as their very birthright—a new thinking seems to be taking hold, one that is increasingly drawing attention away from readings—of scores or meanings that are the result of acts of inscription—and focusing it on the materiality of musical communication, issues of sensuality, and the like.
>
> (p. 2)

It is important, however, to recognize that there is also recent work in ethnomusicology and musicology that specifically focuses on questions of sound, an assertion that Erlmann (2004, p. 2) makes directly after the material quoted above, can be seen in such places as the special interest group for sound studies in the society of ethnomusicology, and is evident in collections about the nature of sound meanings (e.g., Brandt, Duffy, & MacKinnon, 2009; Bull & Back, 2003a; Drobnick, 2004). In light of the doubled complexity of the field, a multiplicity of perspectives about the sonic from a multitude of disciplines, my purpose here is to provide a flavor of sound studies through a brief discussion of four prominent conceptualizations in the field: soundscape, acoustemology, sound art, and sound/methodology.[4]

Soundscape

Soundscape (also referred to as acoustic ecology) is a term and field that originated with composer, educator, and theorist R. Murray Schafer (1977) and developed with an associated group of researchers (e.g., Truax, 1978) at the World Soundscape Project at Simon Fraser University. Schafer envisioned soundscapes as a means for rendering all acoustic environments as music and anything that makes sound as musicians. "Behold the new orchestra: The sonic universe! And the musicians: anyone or anything that sounds" (p. 5).

For Schafer, this understanding has a strong ethical component. Because music created by a particular culture/society both reflects and is informed by that group's understandings of sounds and music-making, it is also related to their sociocultural norms and values (1977, p. 7). This parallels a central tenet on which contemporary ethnomusicology rests, the overarching idea that music is a social construction that can be used as a means to interpret sociocultural ways of knowing.

Just as an analysis of landscapes can provide important information about the contours, contexts, and histories of a particular environment, collecting and analyzing sound can similarly enhance our understandings of a given terrain and the contexts that inform that environment.

> The soundscape is any acoustic field of study. We may speak of a musical composition as a soundscape, or a radio program as a soundscape or an acoustic environment as a soundscape. We can isolate an acoustic environment as a field of study just as we can study the characteristics of a given landscape.
>
> (Schafer, p. 7)

Due in no small part to its broad definition, "soundscape" has been utilized as a means to regard meanings in a wide variety of contexts, ideas, and histories (Born, 2015; for a more detailed recent review of the term, check Kelman, 2010).

Yet, soundscape as defined by Schafer "is deeply problematic as a central figure for sound studies" (Eisenberg, 2015, p. 198). Eisenberg continues:

> Not only is it grounded in normative ideas of which sounds "matter" and which do not, it groans under the weight of the irony that it is born of the very modern technologies of sound productions that Schafer decries as sources of "lo-fi" "pollution" (Helmreich, 2010). Even the term's greatest strength—the fact that it "evokes a whole complex set of ideas, preferences, practices, scientific properties, legal frameworks, social orders, and sounds"—is also a weakness insofar as it diminishes the term's heuristic value.
>
> (Eisenberg, 2015, p. 198)

Further to this point, much of the work on soundscape ironically misses the field of critical geography, a point that is clear when reading Eisenberg's work as compared to many colleagues' discussions at the intersections of sound/space/place.

Acoustemology

Acoustemology is a term associated with Stephen Feld, a central figure in ethnomusicology and sound studies. Feld's (1982) development of acoustemology began with his work with people of the Bosavi region in Papua New Guinea. Starting with Schafer's concept of soundscape, he began to develop, "the idea of an ethnography of sound, or study of sound as a symbolic system, a cultural system, in order to relate the importance of acoustic ecology, particularly the avian rainforest soundscape, to the musicality and poetics of Bosavi laments and vocal song" (Feld, 2000, p. 183). This, in turn, led to the development of what Feld calls "acoustemology," a move he sees as not only a "natural development" in his scholarship but also a "step in critical response to research in acoustic ecology that artificially separates sonic environments from the pervasiveness of human experience" (Feld, 2000, p. 184). Acoustemology "means an exploration of sonic sensibilities, specifically of ways in which sound is central to making sense, to knowing, to experiential truth" (Feld, 1996, p. 97).

In sum, "by acoustemology I wish to suggest a union of acoustics and epistemology, and to investigate the primacy of sound as a modality of knowing and being in the world" (Feld, 2000, p. 184). Parallel to Schwab's (1969) notion of "curriculum-in-action" (p. 12), what we now often refer to as enacted curriculum, Feld (2015) writes that "acoustemology joins acoustics to epistemology to investigate sounding and listening as a knowing-in-action: a knowing-with and knowing-through the audible" (p. 12), understandings that he links to both relational ontology and to Dewey's (1960) use of hyphens in "inter-action" and "trans-action" to emphasize their "across-ness and between-ness" (p. 13).

Sound Art

With roots in fields such as musique concrète, the mid-20th century Western art music of composers such as John Cage, and early recording techniques associated with the development of radio and sound recordings (analog to digital), the category of "sound art" is a relatively recent development (Kim-Cohen, 2009, p. xix). As its name implies, sound art is artwork that is comprised of sound, focuses on sound, or in which sound is a key element (LaBelle, 2007; Licht, 2007; Kahn, 2001; Kim-Cohen, 2009).[5] Because of this, sound art is a relatively open category in which a sculpture with MIDI trigger pads for sounds that activate when touched, a room empty except for a speaker, and, if Kim-Cohen's trajectory of a non-cochlear sound art continues to resonate in the field, words about sound-as-art, can all be labelled sound art. While sound art is not necessarily confined to galleries or allocated public spaces, some of its strength as a medium indeed lies in the ways in which sounds cannot be contained by physical objects. Such art spaces continue to be two of the central ways in which sound art is presented—the sounds of laughter that accompanied riders on the escalator to the exhibition hall at the 2010 meeting of the AERA in Denver is but one example. Thinking about how sounds can be organized in this fashion and in associated art forms such as radio has also informed questions about how one might conduct research in and through sound.

More Contemporary Sound Methodologies and Theories

Furthermore, there is emerging scholarship in research methodology and sound theory that has developed in conversation with the consideration, practices, and creation of soundscapes, acoustemology, and sound art. Where Bauer (2000) focuses on "music and noise" as sources for qualitative data, Drever (2002) works to marry ethnography and soundscape for the purposes of music composition. Similarly, Makagon & Neumann (2009) note the possibilities for and relationship between "audio documentary and the ethnographic experience" in the processes of "recording culture." Along these lines, Jean-Paul Thibaud's theorizing "towards a praxiology of sound development" (www.sensorystudies.org/?page_id=358), in which he suggests that "it is necessary to challenge three main socially recognized categories of sound: music, speech, and noise" and that "the pragmatic dimension of the acoustic environment has also been largely under-estimated until now" (np).[6]

Additionally, Samuels, Meintjes, Ochoa, & Porcello (2010) offer a very thorough and thoughtful work on the possibilities at the intersection of ethnography and soundscape. My own work along similar lines of thought seeks to further such scholarship by further situating the intersection of ethnography and sonic representation between both previously suggested methodological possibilities and the larger field of interpretive research practices, and to put such theoretical possibilities into empirical practice (Gershon, 2011a, 2011b, 2013; Gershon and the Listening to the Sounds of Science Project, 2012). Finally, it is important to reiterate that, like sound art, these concepts arose from developments in sound recording, music composition, and conceptualizations of sounds, noise, and silence from the previous century to the present.

What this scholarship shares is an attention to the ways in which sounds, any and all sounds, are meaningful to the hearer/listener. The sonic is also a part of all ecologies, regardless of their definition or contours, and doesn't require human presence to be meaningful. Along similar lines, sounds do not need to be compared to other senses or previous interpretations and experiences to have value. Noise is never simply noise but is instead some kind of sound that conveys socioculturally contextualized and embedded knowledge both about the source of the sound and its place in relation to the listener, both literally and metaphorically. In much the same way, a recorded sound is necessarily never the same as a sound event but is instead a version of that moment in time as recorded by someone using a specific set of recording instruments with all the biases, norms, and values inherent to the ways in which local actors make sense.

This intentioned focus on sound has also yielded another two central sets of understandings regarding Western conceptualizations of how information and knowledge are received and conceptualized. The first of these sets of understandings are examinations of the intended and unintended results of a primacy of vision and the eye within in Western knowledge from more historical constructions to contemporary theory and philosophy (see Howes, 2003; Ihde, 2007; O'Callaghan, 2007). "[V]isual information occupies a privileged epistemic role, and our language frequently reflects a tight coupling of seeing with knowing. We evaluate *views*, have *insights*, and *see* what is at issue" (Nudds & O'Callgahan, 2009, p. 1, italics in original).

As Erlmann (2010) describes, "the ear troubles some of our most entrenched clichés" (p. 24)—visual metaphors for ontological, theoretical, and philosophical understandings of how people make meaning immediately fall short when applied to sonic and other non-ocular, sensual meaning-making. The assumption that non-ocular senses are identical to vision's functions and relationships to understanding lead to both missed theoretical opportunities and necessarily false constructions of meaning (Bull & Back, 2003b; Ihde, 2007; O'Callaghan, 2007). Consider the following: "[t]he blink of an eye lasts three hundred milliseconds. The blink of an ear lasts considerably longer. From birth to death the ear never closes" (Kim-Cohen, 2009, p. xvii).

Second, an understanding of the primacy of vision and the categorization of the senses into five discreet categories is a distinctly Western construction that has a history of reifying notions of Western supremacy over "other" non-Western ways of knowing that in many ways continues unabated, even in studies of the senses or the visual and performing arts (e.g., Nooshin, 2003). As anthropologists have demonstrated in recent decades (e.g., Geurts, 2002; Stoller, 1997; for a review, see Howes, 2003), non-Western constructions of meaning do not necessarily place mind over body or vision over sound over taste, touch, and smell. Bringing such constructions of knowledge to non-Western contexts can and often does lead to misunderstandings of local norms and values and, in turn, to interpretations that reveal more about the researcher than represent those researched (e.g., Fox & King, 2002; Faubian & Marcus, 2009). Furthermore, as Erlmann (2010) intricately documents, even a Cartesian splitting of mind and body, of valuing sight over sound, overlooks deep connections between reason and resonance that echo throughout modernity.

Thus, sound studies articulates the meaning-full-ness of sound, its connection to the human experience (profound, mundane, and otherwise), and reminds us of the inclusive nature of sound, both in its perception and the sources that trigger reception. In these ways, the sounds we hear every day are central to our understandings of how we know ourselves, our relationship to our environments, and the people, animals, ecological life, and "inanimate" material/objects (rocks on a path, bulldozers on the freeway) that populate our local and less local ecologies. Sounds contain meanings that are literally embodied, absorbed both through our orifices and throughout our systems (Erlmann, 2010)—the louder the volume, the more this point literally resonates and I will return to it in the following chapter.

On one hand, the sound ideas and ideals as conceptualized in sound studies should ring familiar with curriculum studies. It is a field that has its own disciplinarity (Pinar, 2007) yet it is also interdisciplinary and critical, as well as the site of multifarious discussions that question dominant notions of knowing while theorizing the possible and pushing towards its practical implications. Additionally, curriculum studies seems poised to contribute towards Kim-Cohen's (2009) notion of a non-cochlear sonic art in which sounds can be textually represented concepts and constructs—an idea rooted in Derrida that Erlmann (2010) implicitly challenges in his support of critiques of Derrida's "phonophobia" and his own concerns about Derrida's "demotion of the ear" (p. 12).

On the other hand, in educational literature both in and out of curriculum studies, sound is often overlooked in favor of music and talk or equated with unwanted noise (Chapter 1). As a result of this orientation there is a continued emphasis on the primacy of vision, texted rather than embodied experience of knowledge in most educational scholarship, and an often uninterrupted assumption that non-visual sensory understandings (literal and metaphoric) can be recognized through the same ideas and practices as those used for vision. This also has implications for the medium through which one attends to sound. From this perspective, it is more the ways in which curriculum studies has utilized text and a lack of attention to sound (as opposed to music and speech) than it is that scholars have written rather than listened to text. At first blush, such omissions might seem either semantic or surface in nature. However, because what is at issue here speaks to the ways in which people make sense of their worlds, a lack of attention to the sonic misses central ways in which knowledge is assembled and disseminated. The following section is dedicated to empirically demonstrating just this, how sounds form embodied educational systems of meaning.

One Grade, Two Contexts, and a Microphone: Sounds as Embodied Educational Systems

From 2008–2012, I worked with Mrs. Grindall[7] and the successive groups of fifth graders she taught. Due to the way in which our collaborative project was conceived, I am able to use her and the students' actual names, a possibility that is helpful here because it means that the sounds in these two educational contexts do not need to be further edited to remove identifying information.

In light of the concepts that are part and parcel of scholarship in sound studies briefly outlined in the previous section, it is important to again note that I am not claiming that a recorded instance of any event either accurately captures that event or can be regarded as the event. This is because myriad interwoven layers of context, not the least of which are time and physical presence, contextualize any sound and because it is necessarily incomplete, as is the case with any interpretation.

This does not mean, however, that one cannot gain embodied knowledge by listening to sound recordings. To the contrary, human beings are excellent at recognizing even incomplete sounds (e.g., Deutsch, 2010; Kassabian, 2013). Therefore, as long as one does not make assumptions about a microphone and/or a recorder truly capturing a sonic environment—overlooking that a microphone's placement, range, and the sounds that cross its elements are the result of the person recording and that the equipment capturing the recording, and, no matter how sophisticated the setup or process, produce some version of invented/interpreted sound—listening to a particular context can indeed convey embodied knowledge. Additionally, as sounds are contextualized by other sounds, it is both the presence of any given sound and their combination with other sounds that educate the listener about herself, others, and her environment.

To these ends, the remainder of this section has been organized as follows. I begin with a brief description of the two contexts in which the sounds below were recorded, an explanation that includes a discussion of the teacher and students involved. I then provide a texted version of the first context, just over a minute of sonic data in which students went from listening on the rug with their teacher to working at their desks in their assigned groups.

Next, I offer an mp3 file (selected for its near universal play-ability rather than sound quality) of the transition so that the listener can hear the sound data used to create the text of the transition and appreciate the differences between the two versions. Finally, I provide another mp3 file of Mrs. Grindall with a different group of students doing an exercise in a forest so that the reader/listener can performatively experience the degree to which sounds form educational systems of meaning across contexts and participants.

Contexts

The two contexts presented here are of Mrs. Grindall and two successive years of fifth grade students, those she had last academic year (2009–2010) and those who began the 2010–2011 academic year as her students. Where fifth graders in the classroom context had been with Mrs. Grindall for just over seven months, students in the second context below had become her students quite recently, having arrived in her room only three weeks prior to the recording. Additionally, while the first recording is of a rather traditional educational context, a classroom in an urban elementary school, the second recording is equally nontraditional. It is a recording of students playing a sensory game in a national forest as part of an annual three-day, overnight trip Mrs. Grindall plans so that "they [her students] can experience science and nature for themselves, something they really almost never get a chance to do and always say is one of the highlights of their year." As one boy who is the youngest of three brothers, all of who have had Mrs. Grindall as their fifth grade teacher, shared with me about his time at the Cuyahoga Valley National Park: "Finally. After all these years it's my turn, and I'm so glad

I'm here. It's as cool as I thought. Kinda better . . . because this time it's me [here]."

The students in both classes are majority African American (16 of 18 students in 2009–2010 and 19 of 22 in 2010–2011, two of whom are students of color who are not Black) and come primarily from poor to working class families who live in Akron, Ohio. Mrs. Grindall is a widely respected teacher with over 35 years of teaching experience. When this recording was conducted (AY 2009–2010), the classroom was organized with a large rug in the middle of the floor on which the students sat to work, listen to stories, and read; there was a SMART Board at one edge of the rug and a rocking chair in which Mrs. Grindall read to students on the other; desks organized so that students sat in groups of four facing one another lined the walls of the classroom outside the rug; and a library filled with books behind another set of rugs where the students could sit or lie down as they read.

Sounds as Educational Systems In and Out of the Classroom

At the point that the transcription begins, students sat on the rug as Mrs. Grindall finished a discussion about courage and moved on to the topic of their next assignment, a group writing project about perseverance. In the transcription "(sec)" is an indicator of lengths of silence between utterances, "(:##)" represents the time elapsed in the sound data strip since it began, an ellipsis ". . ." is used to indicate interrupted speech, and a bracket "]" is used in combination with indention to indicate lines of overlapping talk.

Mrs. Grindall:	Alright? Perseverance is written in the center box already. (5 sec) And you're going to do looots of sharing with each other. Nobody's asking you to do, as I usually don't do, I want you to put your brains together, I want you to put your brains together. So perseverance is to keep trying and trying different things until you succeed. (:26) So as soon as you are ready, you are gonna get where your group meets, okay?. . . .]
Student 1:	Mrs. Grindall?]
Mrs. Grindall:	where your reading group meets, your clipboard. . .]
	(Student conversation unintelligible)
Mrs. Grindall:	(:42) and I can't wait to see your *awesome* sharing. I see Karis is already gathering her group and going to where they normally meet.]
Student 2:	Mrs. Grindall, I need a sample [sheet]]
Mrs. Grindall:	(quietly to student): The blue one too?]
Student 2:	Yeah

Mrs. Grindall: [louder voice to the whole class] Take the blue one with you so you have a sample, that's what that is. It's a different *theme* but it will give you a sample. It'll give you an, "Oh, this is what this means" (2 sec, 1:06). I'll come around and sit in with your group for a while. I wanna listen to you.

I want to listen to your questions. I wanna help you out.
(Overlapping student conversations for 4 seconds as students begin to work)

While there are certainly more in-depth ways of transcribing talk-and-sound-as-text (e.g., Erickson, 2004; van Leeuwen, 1999), the above transcription also carries some sound markers that are not as frequently used. For example, in addition to noting overlapping talk, italicizing words for emphasis, and the insertion of time markers, the transcription of talk above also notes the number of seconds in wait time between utterances and stage direction-like presentations about the quality of talk (e.g., "quietly to student"). Brackets for overlapping dialogue, while not necessarily unusual, are also less commonly utilized.

Transcriptions of talk of this kind are so ubiquitous in qualitative scholarship that they have become an expected part of reporting data. While we often give notice that all such work is necessarily a transcription and have reached a place in the history of social science that the partiality of representation is equally commonplace, we, members of the qualitative research community, rarely move beyond such details of description, if such details are included at all. Instead, we have come to accept such texted talk as adequately conveying sonic data. With this thought firmly in mind, here, then, is the minute and 18 seconds of sonic data used to create the above transcription.

ClassroomSounds.mp3

Most expressly missing from the sound sample are all the classroom background sounds (from the "bell ringing" that Mrs. Grindall speaks through to the increased rustling of papers), student movement, and conversation that are present to some degree throughout Mrs. Grindall's talk, sounds that markedly increase as students stand to go work with their groups. These sounds note students' active presence in the room in a way that texted transcription does not. This is no minor difference as in the texted version it appears as though the teacher is working in a relative vacuum.

Of equal importance, the seemingly unorganized sounds that are not notated, and would be quite difficult to notate in their complexity in any system of transcription, combine to form another layer of meaning, as informative as the talk transcribed above. This point can take a moment of mental sonic realignment as it is an inversion of how sounds are conventionally perceived. Imagine that this recording was not a recording of a classroom through a pair of small condenser microphones set to capture the classroom as widely as possible but was instead assembled in a sound studio so that every sound was discreetly recorded and could therefore be manipulated individually.

In this imagined scenario, one could then lower the volume on all of the talk until only the "sounds" of the classroom remain. These sounds include the rustling of papers, scraping of chairs, shuffling of feet, and, depending on one's conception of how this might be actualized, the increasing number and volume of students' simultaneous conversation that becomes the complex sound we recognize as the murmur or buzz of conversation—a kind of inversion of most Foley artists' work in film. To my mind, this last aspect would be included as the murmur of students in classrooms is qualitatively different from the murmur of a crowd in a restaurant or, as will be demonstrated momentarily, students talking and walking in pairs in the forest.

Furthermore, the meanings ascribed to these sounds are socioculturally constructed and contextualized. The loud buzzer that interrupts Mrs. Grindall's talk is but one example of this understanding. In another context, this buzzer might have one of many other possible meanings, from a warning before a gate is opened to something closer in meaning such as the change of a shift at a factory. In the US classroom context of this recording, this electronic tone is referred to and functions as a "bell" by school actors, even though it does not have the tonal qualities associated with bells in general or school bells in particular. This sound is of interest to my argument here for yet another reason. Although it is neither talk nor music, it is a sound intentionally organized by the school as a sonic marker (Schafer, 1977) significant to local actors, in this case the "bell" for younger students to go to lunch.

Additionally, even though no such sounds are noticeably present on this recording, it was not unusual for city sounds to enter or vibrate the classroom's windows. These sounds included passing sirens, car horns, and the loudspeakers from the car lot around the corner from the school. Sounds often associated with less populated ecosystems also impacted the classroom throughout the year. Intense rainstorms lashed at the windows, thunder rolled, and students heard the way a soundscape changes at the first snow of the year even on a busy city street.

Accordingly, sounds that are most often disregarded as either negatively interrupting learning environments or as not meaningful are indeed *educational* in that they convey meaning to the listener and *systemic* in that the listener places those meanings into conceptual categories that inform their understandings. The same holds true for what I have referred to here as organized sounds, talk and music. In addition, this knowledge is literally embodied, resonating within one's body regardless of one's ability to hear them. Where a blind person's inability to see is equivalent to a loss of vision, a profoundly deaf person is still able to dance in time and play music due to the ways in which sounds resonate within the body as is the case with profoundly deaf percussionist/composer Dame Evelyn Glennie (www.evelyn.co.uk/Evelyn_old/live/hearing_%20essay.htm).

Thus, all sounds form educational systems. They tell us about our environment and our relationship to others and reveal as much about how we understand the world as they convey meanings to us as listeners. I offer the following sonic strip of data in order to demonstrate that these understandings are not particular to this context but are instead a way of conceptualizing sounds and sonic data. In it, that year's (2010–2011) group of fifth graders was in the middle of an exercise in the Cuyahoga Valley National Park that their guide for the morning called a "sensory walk."

For this exercise, one student (A) was to use her or his voice to guide a blindfolded partner (B) to a tree somewhere off the trail on which they had been walking. The blindfolded partner was tasked with "investigating the tree using all your other senses, touch it, smell it, see how it sounds." The blindfolded partner was then led back to the trail at which point the partners *swapped* the blindfold and the *originally* blindfolded partner (B) guided their now-blindfolded peer (A) back to the tree she or he thought they recognized based on the non-sighted sensory investigation that person conducted while blindfolded. This strip of sonic data follows some students as they began their sensory walk.

SensoryWalk.mp3

In the sonic information, one can hear students guiding other students through the woods, one student telling another to step "right here, right here, right here"; the depth and breadth of an educational space defined by the crunching of leaves and the distance one can hear the instructor ("that's far enough") rather than real or imagined physical boundaries such as classrooms and hallways; the imperfections of the recording as I walked, the rustling of the cord, and interruptions of wind. As can be heard here, sounds do indeed form educational systems—meanings participants constructed as they engaged in their learning activities in the classroom and forest, meanings I made by following a particular set of choices while walking with an audio recorder through the woods with the class, and understandings you have built as a listener to these events.

As was the case in the classroom, sounds informed students about themselves and their contexts. Of equal importance, these two sonic data strips also function performatively, educating the listener-reader of this piece about these two educational contexts. They are nested, layered sets of sonic interpretations in which 1) sounds form educational systems for the listener of the recorded context in order that the listener might contemplate 2) the nature of all sounds as educational systems for students as 3) they form their own meaning systems based on the local sounds that resonated in their person at that time (in the classroom and forest). Thus, all sounds form educational systems of meaning, embodied ways of making sense (Ellsworth, 2005;

Erlmann, 2010) of our place in the world and our lived environments on a moment-by-moment basis.

Sound Implications: Deep Listening as Education

An argument for sounds as educational systems of knowledge has some not insignificant implications for approaches and conceptualizations of educational ways of beingknowingdoing. If all sounds are meaningful (e.g., Back, 2007; LaBelle, 2010; Feld, 2005), then what it means to engage in educational interactions and processes—the transferring of ideas and ideals that often occurs between people, but is by no means limited to human-to-human interactions or to forms of schooling—then a central aspect of anything that might be called "education" must be listening.

Being a working musician is a life of study. We study those who came before us and after, genres and their associated conventions, ideas and ideals about what it means to make music, our own practice literally and performatively through processes of practice, the forces and dynamics of sound, the possibilities and limitations of our particular bodies to make sounds through an instrument (including our bodies), compositions in many formats, traditions and histories, and so on.

The same can be said in many ways of educators. Yet, there is a significant and deeply ironic difference. Although sonic communication is central to both professions, where teachers most often talk, musicians must listen to others. Even the most self-centered, self-involved musician, the top of the narcissistic heap, must truly listen to others, because to engage in the musical process is to feel in, through, and with sound. By this I don't mean that teachers don't listen or engage with children or attend to myriad things in the world around them. But, for a moment, just do this. Close your eyes and imagine a teacher engaged in the act of teaching. What are they doing? Okay, perhaps before you read this passage where I was rather heavy handed with the lede. As but one cultural artifact, consider the teacher sounds in Charlie Brown specials: "mwah, mwahwahwah, mwhaaa," the muted trombone drones on.

This is likely because we have come to a place in education, or may have always already been here, where we believe that answers are the key to a strong education. We present education as often-discreet-eventually-connected skills that are given to students, then measure that education according to the degree to which students arrive at predetermined answers, often through rather prescriptive steps. As a process, schooling is often working through questions someone else has asked through to answers someone else has provided. This tends to be the case even when students are engaged in so-called inquiry-based education; we say to students, here are the steps in the inquiry process. Problem-based education is often the same thing: here's a problem, you figure out how you reach it, then we check to see if it's feasible. The shame of it all is that education, in or out of schools, is about wonder—and at the heart of wonder is questions. To speak of sounds of as educational systems, an understanding that all sounds carry meaning, is to re-remember that deeply listening to students resides at the core of being-knowingdoing education.

Part of the reason it might be difficult to conceive of sounds as educational systems resides not only in how sound has been conceptualized but also due to the ways in which thinking about education is often bounded. In the same way that certain tasks have come to be regarded as on or off task according to a teacher's construction of a lesson, it seems as though sounds have been implicitly theorized in a similar fashion: Sounds that are part of the teacher's agenda are educational, those that are not, aren't. This is not to say that sounds cannot be distracting or detrimental to learning but rather that there appears to be a tendency to turn to outside sounds as distracters, loud air conditioning units and nearby traffic for example, rather than those sounds generated by teachers or other adults—wearing squeaky shoes during a test or consistently giving help to a favored student *sotto voce* so that only a directed few or one benefit from the extra information, are but two examples.

A lack of theorization of sound education is also interrelated to the ways in which the senses have been undervalued and underutilized as sources of knowledge both in and out of education. A sonic approach to education might harken to questions of experience and meaning that lie at the center of curriculum studies. For example, what might be possible if one were to use sound metaphors to consider schooling rather than "frames" or "lenses?" (for more on this point, Chapter 3). Given that non-Western constructions of sound do not necessarily prescribe to either a "five senses" model or a primacy of vision, turning to such understandings may well benefit curriculum scholars and prove to be useful for such fields as post-colonial studies and indigenous studies and methodologies.

Finally, regardless of how they are conceptualized, the senses are utilized in concert with one another. Therefore, attention to one sense is necessarily attention to other senses, a question of foregrounding rather than of primacy. As stated in the introduction, my intention here has been to demonstrate that sounds form educational systems of embodied meaning, not to elevate sound over sight but to place the sonic on par with the visual. An opportunity, to paraphrase students on their sensory walks, to smell, touch, taste, and listen, rather than simply see the trees for the forest and the forest for the trees.

Notes

1 By embodied knowledge I mean the multitude of possibilities that sounds intersect with people and the similarly always complex ways that anything might be considered either "knowledge" or "knowing." Knowledge, from this construction, is not limited to ideas known in a head or brain, falsely split from a body or feelings, but is instead any pathway or possibility where a person gains understanding of some sort. That they are embodied is a reminder that knowledge is always embodied, even in a more

classical conceptualization of a Cartesian split, the head and/or brain is necessarily part of a body.

2 It is important to note that sound is more than a social construction; sounds are an integral aspect of being human for the deaf and hearing alike as well as physical phenomena that exist regardless of human perception or presence. Although this chapter focuses on the meanings ascribed to sounds, overlooking the physiological for the social can lead to both missed understandings about the nature human understandings (Erlmann, 2010), the ways in which deaf people are marginalized (Lane, 1999), and the politics surrounding questions of deafness and hearing (e.g., Valente, Bahan, and Bauman, 2011) that will be covered in greater detail, though by no means exhaustively, in Chapter 4.

3 Nina Sun Eidsheim (2015), *Sensing Sound* and David Novak and Matt Sakakeeny's edited book (2015) *Keywords in Sound* on sound sensibilities and Elizabeth Ellsworth (2005) *Places of Learning: Media, Architecture, Pedagogy* and Eve Kosofsky Sedgwick (2003) *Touching Feeling: Affect, Pedagogy, Performativity* on making educational sense.

4 For a more in depth discussion of terms central to sound studies, please find *Keywords in Sound*, also suggested in the previous note; for an overview of the kinds of scholarship that comprise the field, please check out Jonathan Sterne's (2012), *Sound Studies Reader* or Trevor Pinch and Karin Bijesterveld's (2013) *The Oxford Handbook of Sound Studies*.

5 Although there are certainly points of contention between those cited here, the ideas presented in this section are congruent between all four works and consistent in the same general fashion with other discussions of sound art.

6 For more on conceptualizing noise and its possibilities, also check Hillel Schwartz's (2004/1998) piece, On Noise, in Mark M. Smith's (2004) edited collection, *Hearing History: A Reader*.

7 Mrs. Grindall has since retired and I remain indebted to her and all the other teachers who participated in this longitudinal collaborative sonic ethnography.

References

Aoki, T. T. (1991). *Sonare* and *vidare*: Questioning the primacy of the eye in curriculum talk. In G. Willis & W. H. Schubert (Eds.), *Reflections from the heart of educational inquiry: Understanding curriculum and teaching through the arts*. (pp.82–89). Albany, NY: State University of New York Press.

Appelbaum, P. (2007). *Children's books for grown-up teachers: Reading and writing curriculum theory*. New York: Routledge.

Attali, J. (1985). *Noise: The political economy of music* (B. Massumi, trans.). Minneapolis, MN: University of Minnesota Press.

Ayers, W. C., & Miller, J. L. (Eds.). (1997). *A light in dark times: The unfinished life of Maxine Greene*. New York: Teachers College Press.

Back, L. (2007). *The art of listening*. Oxford: Berg Press.

Bauer, M. W. (2000). Analysing noise and music as social data. In M. W. Bauer & G. Gaskell (Eds.), *Qualitative researching with text, image, and sound* (pp. 263–281). London: Sage Press.

Bellack, A. A., Kliebard, H. E., Hyman, R. T., & Smith Jr., F. L. (1978). *The language of the classroom*. New York: Teachers College Press.

Born, G. (Ed.). (2015). *Music, sound and space: Transformations of public and private experience*. Cambridge, UK: Cambridge University Press.

Brandt, R., Duffy, M., & MacKinnon, D. (Eds.). (2009). *Hearing places: Sound, place, time and culture*. Cambridge, England: Cambridge Scholars Press.

Bull, M. (2001). Soundscapes of the car: A critical study of automobile habitation. In D. Miller (Ed.), *Car cultures* (pp. 185–202). Oxford: Berg Press.

Bull, M., & Back, L. (2003a). *The auditory culture reader*. Oxford, UK: Berg.

Bull, M., & Back, L. (2003b). Introduction: Into sound. In M. Bull & L. Back (Eds.), *The auditory culture reader* (pp. 1–24). Oxford, UK: Berg.

Clifford, J. (1986). Introduction: Partial truths. In J. Clifford & G. Marcus (Eds.), *Writing culture: The poetics and politics of ethnography* (pp. 1–26). Berkeley: University of California Press.

Clough, P.T., & Halley, J. (2007). *The affective turn: Theorizing the social*. Durham, NC: Duke University Press.

Deleuze, G., & Guattari, F. (1987). *A thousand plateaus: Capitalism and schizophrenia*. Minneapolis, MN: University of Minnesota Press.

Deutsch, D. (2010, July/August). Speaking in tones. *Scientific American Mind*, 36–43.

Dewey, J. (1960). *On experience, nature and freedom*. New York: Bobbs Merrill.

Dimitriadis, G. (2009). *Performing identity/performing culture: Hip hop as text, pedagogy, and lived practice* (Revised ed.). New York: Peter Lang.

Drever, J. L. (2002). Soundscape composition: The convergence of ethnography and acoustamatic music. *Organised Sound*, 7(1), 21–27.

Drobnick, J. (Ed.). (2004). *Aural cultures*. Toronto, ON: YYZ Books.

Eidsheim, N. S. (2015). *Sensing sound: Singing and listening as vibrational practice*. Durham, NC: Duke University Press.

Eisenberg, A. J. (2015). Space. In D. Novak & M. Sakakeeny (Eds.), *Keywords in sound* (pp. 193–207). Durham, NC: Duke University Press.

Eisner, E.W. (1977). On the uses of educational connoisseurship and criticism for evaluating classroom life. *Teachers College Record*, *78*(3), 345–358.

Ellsworth, E. A. (2005). *Places of learning: Media, architecture, pedagogy*. New York: RoutledgeFalmer.

Erickson, F. (1982). Classroom discourse as improvisation: Relationships between academic task structure and social participation structure in lessons. In L. C. Wilkinson (Ed.), *Communicating in the classroom* (pp. 153–181). New York: Academic Press.

Erickson, F. (2003). Some notes on musicality and speech. In D. Tannen & J. E. Atalis (Eds.), *Linguistics, language and the real world: Discourse and beyond (Georgetown University: Round table on languages and linguistics)*. Washington, DC: Georgetown University Press.

Erickson, F. (2004). *Talk and social theory: Ecologies of speaking and learning in everyday life*. Malden, MA: Polity Press.

Erlmann, V. (Ed.). (2004). *Hearing cultures: Essays on sound, listening and modernity*. Oxford: Berg Press.

Erlmann, V. (2010). *Reason and tesonance: A history of modern aurality*. Cambridge, MA: Zone Books.

Faubian, J. D., & Marcus, G. E. (2009). *Fieldwork is not what it used to be: Learning anthropology's method in a time of transition*. Ithaca, NY: Cornell University Press.

Feld, S. (1982). *Sound and sentiment: Birds, weeping, poetics, and song in Kaluli expression*. Philadelphia: University of Pennsylvania Press.

Feld, S. (1991). Sound as a symbolic system: The Kaluli drum. In D. Howes (Ed.), *The varieties of sensory experience. A sourcebook in the anthropology of the senses* (pp. 79–99). Toronto: University of Toronto Press.

Feld, S. (1996). Waterfalls of song: An acoustemology of place resounding in Bosavi, Papua New Guinea. In S. Feld & K. H. Basso (Eds.), *Senses of place* (pp. 91–135). Santa Fe, NM: School of American Research Press.

Feld, S. (2000). Sound worlds. In P. Kruth & H. Stobart (Eds.), *Sound (The Darwin College Lectures)* (pp. 173–200). Cambridge, UK: Cambridge University Press.

Feld, S. (2005). Places sensed, senses placed: Toward a sensuous epistemology of environments. In D. Howes (Ed.), *Empire of the senses: The sensual culture reader* (pp. 179–191). Oxford: Berg Press.

Feld, S. (2015). Acoustemology. In D. Novak & M. Sakakeeny (Eds.), *Keywords in sound* (pp. 12–21). Durham, NC: Duke University Press.

Fox, R. G., & King, B. J. (Eds.). (2002). *Anthropology beyond culture*. Oxford, UK: Berg.

Gandy, M., & Nilsen, B. (Eds.). (2014). *The acoustic city*. Berlin: Jovis Verlag.

Geertz, C. (1983). *Local knowledge: Further essays in interpretive anthropology*. New York: Basic.

Gershon, W. S. (2006). Collective improvisation: A theoretical lens for classroom observation. *Journal of Curriculum and Pedagogy, 3*(1), 104–136.

Gershon, W. S. (2011a, January 23). Sonic ethnography: Sound meanings and meaningful sounds. Paper presented at the Narrative, Arts-Based, & "Post" Approaches to Social Research Conference (NAPAR), Tempe, AZ.

Gershon, W. S. (2011b, April 9). Innovative approaches to scholarship II: Listening to the sounds of science: Knowledge and making sense in urban middle grades classrooms. Sound Installation/Data Representation. *American Educational Research Association*. New Orleans, LA.

Gershon, W. S. (2013). Vibrational affect: Sound theory and practice in qualitative research. *Cultural Studies ←→ Critical Methodologies, 13*(4), 257–262.

Gershon, W. S., and The Listening to the Sounds of Science Project. (2012, March 17–July 15). *Making sense of science: The sounds of teaching and studenting in four urban classrooms*. [Art/Sound Installation]. Akron Art Museum.

Geurts, K. L. (2002). *Culture and the senses: Bodily ways of knowing in an African community*. Berkeley, CA: University of California Press.

Greene, M. (1995). *Releasing the imagination: Essays on education, art, and social change*. San Francisco: Jossey-Bass.

Gregg, M., & Seigworth, G. J. (Eds.). (2010). *The affect theory reader*. Durham, NC: Duke University Press.

Gustafson, R. I. (2009). *Race and curriculum: Music in childhood education*. New York: Palgrave Macmillan.

Heath, S. B. (1983). *Ways with words: Language, life, and work in communities and classrooms*. Cambridge, UK: University of Cambridge Press.

Helmreich, S. (2007). An anthropologist under water: Immersive soundscapes, submarine cyborgs, and transductive ethnography. *American Ethnologist, 34*(4), 621–641. DOI: 10.1525/ae.2007.34.4.621.

Howes, D. (2003). *Sensual relations: Engaging the senses in culture & social theory*. Ann Arbor, MI: University of Michigan Press.

Howes, D. (Ed.). (2005). *Empire of the senses: The sensual culture reader* (pp. 179–191). Oxford: Berg Press.

Ihde, D. (2007). *Listening and voice: Phenomenologies of sound* (2nd ed.). Albany, NY: State University of New York Press.

Jardine, D. (2004). A bell ringing in the empty sky. In W. F. Pinar (Ed.), *Contemporary curriculum discourses: Twenty years of JCT* (pp. 262–277). New York: Peter Lang.

Kahn, D. (2001). *Noise, water, meat: A history of sound in the arts*. Cambridge, MA: MIT Press.

Kahn, D. (2013). *Earth sound earth signal: Energies and earth magnitudes in the arts*. Berkeley, CA: University of California Press.

Kassabian, A. (2013). *Ubiquitous listening: Affect, attention, and distributed subjectivity: Affect, attention, and distributed subjectivity*. Berkeley, CA: University of California Press.

Kassabian, A. (2015). *Ubiquitous listening: Affect, attention, and distributed subjectivity*. Berkeley, CA: University of California Press.

Kelman, A. Y. (2010). Rethinking the soundscape: A critical genealogy of a key term in sound studies. *Senses & Society, 5*(2), 212–234. doi: 10.2752/174589210X12668381452845

Kim-Cohen, S. (2009). *In the blink of an ear: Towards a non-cochlear sonic arts*. New York: Continuum.

LaBelle, B. (2007). *Background noise: Perspectives in sound art*. New York: Continuum.

LaBelle, B. (2010). *Acoustic territories: Sound culture and everyday life*. New York: Continuum.

Landy, L. (2007). *Understanding the art of sound organization*. Cambridge, MA: MIT Press.

Lane, H. (1999). *The mask of benevolence: Disabling the deaf community* (2nd ed.). San Diego, CA: DawnSign Press.

Leeuwen, T. van (1999). *Speech, music, sound*. London: Palgrave Macmillan.

Licht, A. (2007). *Sound art: Beyond music, between categories*. New York: Rizzoli.

McCarthy, C., Hudak, G., Miklaucic, S., & Saukko, P. (Eds.). (1999). *Sound identities: Popular music and the cultural politics of education*. New York: Peter Lang.

Makagon, D. G., & Neumann, M. (2009). *Recording culture: Audio documentary and the ethnographic experience*. Thousand Oaks, CA: Sage Press.

Malewski, E. (Ed.). (2009). *Curriculum studies handbook: The next moment*. New York: Routledge.

Massumi, B. (2015). *The politics of affect*. Cambridge, UK: Polity.

Mehan, H. (1979). *Learning lessons: Social organization in the classroom*. Cambridge, MA: Harvard University Press.

Morris, M. (1999). Music as curriculum text. In M. Morris, M. A. Doll & W. F. Pinar (Eds.), *How we work* (pp. 11–18). New York: Peter Lang.

Nooshin, L. (2003). Improvisation as "other": Creativity, knowledge and power—the case of Iranian classical music. *Journal of the Royal Musical Association, 128*, 242–296.

Nudds, M., & O'Callaghan, C. (2009). Introduction: The philosophy of sounds and auditory perception. In M. Nudds & C. O'Callaghan (Eds.), *Sounds and perception: New Philosophical essays*. (pp. 1–25). Oxford, UK: Oxford University Press.

O'Callaghan, C. (2007). *Sounds: A philosophical theory*. Oxford, UK: Oxford University Press.

Peters, M., Marshall, J., & Fitzsimons, P. (1999). Poststructualism and curriculum theory: Neo-liberalism, the information economy, and the crisis of cultural authority. *Journal of Curriculum Theorizing, 15*(2), 111–127.

Pinar, W. F. (2004). *What is curriculum theory*? Mahwah, NJ: Lawrence Erlbaum Associates.

Pinar, W. F. (2007). *Intellectual advancement through disciplinarity: Verticality and horizontality in curriculum studies*. Rotterdam, NL: Sense.

Pinar, W. F., & Irwin, R. L. (2005). *Curriculum in a new key: The collected works of Ted T. Aoki*. Mahwah, NJ: Lawrence Erlbaum.

Pinch, T., & Bijsterveld, K. (2013). *The Oxford handbook of sound studies*. Oxford, UK: Oxford University Press.

Sakakeeney, M. (2013). *Roll with It: Brass bands in the streets of New Orleans*. Durham, NC: Duke University Press Books.

Samuels, D. W., Meintjes, L., Ochoa, A. M., & Porcello, T. (2010). Soundscapes: Toward a sounded anthropology. *Annual Review of Anthropology, 39*, 329–345.

Schafer, R. M. (1977). *The tuning of the world*. New York: Knopf.

Schippers, H., & Grant, C. (2016). Approaching music cultures as ecyosystems: A dynnaomic model for understanding and supporting sustainability (pp. 303–352). In H. Schippers & C. Grant (Eds.), *Sustainable futures for music cultures: An ecological perspective*. Oxford, UK: Oxford University Press.

Schwab, J. J. (1969). The practical: A language for curriculum. *School Review*, *78*(1), 1–23.

Schwartz, H. (2004/1998). On noise. In M. M. Smith (Ed.), *Hearing history: A reader* (pp. 51–53). Athens, GA: University of Georgia Press.

Sedgwick, E. K. (2003). *Touching feeling: Affect, pedagogy, performativity*. Durham, NC: Duke University Press.

Shield, B. M., & Dockrell, J. E. (2003). The effects of noise on children at school: A review. *Building Acoustics*, *10*(2), 97–116.

Slattery, P. (2006). *Curriculum development in the postmodern era* (2nd ed.). New York: Routledge.

Smith, M. M. (Ed.). (2004). *Hearing history: A reader*. Athens, GA: University of Georgia Press.

Spindler, G. (Ed.). (1982). *Doing the ethnography of schooling: Educational anthropology in action* (pp. 20–47). New York: Holt, Rinehart, & Winston.

Sterne, J. (2012). *The sound studies reader*. New York: Routledge.

Stewart, K. (2007). *Ordinary affects*. Durham, NC: Duke University Press.

Stoller, P. (1997). *Sensuous scholarship*. Philadelphia: University of Pennsylvania Press.

Thibaud, J-P. *Towards a praxeology of sound environment*. Retrieved from www.sensorystudies.org/?page_id=358

Truax, B. (1978). *The World Soundscape Project's handbook for acoustic ecology*. Vancouver, BC: A.R.C. Press.

Voegelin, S. (2010). *Listening to noise and silence: Towards a philosophy of sound art*. New York: Continuum.

3 Resonance, Reverberation, and Scale

Towards a Sound Philosophy of Education

Ideas and ideals echo throughout the ages: vibrating, viscous waves of possibility that slip through the cracks of supposedly silenced rooms as readily as they careen off one another in a never-ending polyphonic cacophony, or whisper over vast open spaces. Making matters even more complex, even conceptually inanimate things are oscillating, as are purportedly dead languages, antiquated assumptions, and similarly dismissed dreams. This never-ending flow of continual motion is as common to physical objects or sociocultural norms and values as it is to what is collapsed here as beingknowdoing—the inexorably imbricated understanding that to be is to know is to do. They are resonances that, when set in motion (which they always already were), are also reverberations. Understanding educational ways of being and knowing as resonance and reverberations provides particular advantages for conceptualizing not only what education might mean but also how it functions in the everyday lives of those involved in educational interactions, the passing of knowledges from one to another.

This chapter articulates a sound philosophy of and for education. It is also a kind of *clave*[1] through which one can approach this book, possibilities that noticeably ripple between chapters or eek between arguments. Bringing arguments made in the previous chapters to bear on educational philosophy, contemporary understandings in the field have a tendency to utilize ocular metaphors, epistemological constructions that provide *frames* for *seeing* the world.

There are at least the following three difficulties with such a reliance on the ocular. First, they are easily interrupted by non-ocular metaphors. While this may seem to be a rather surface concern, it points to a deep philosophical set of problems. For example, what happens when the *augenblick*, a construct around which much philosophizing has been accomplished, is pressed slightly to become a "blink of an ear" (Kim-Cohen, 2009)? This move not only questions the veracity of ocular-centric constructions from blinks to texts—for what does it say about an idea if it falls apart by shifting senses—but it also presses at the feasibility of framing, its inherently bounded nature as well as the large number of potentialities excluded by such a gaze.

Second, the sonic provides another avenue for wonder. Sound is omnipresent and questions of hearing or listening are as much about physical anatomy as they concern sociocultural constructs about what sounds can mean. Unlike sight that is necessarily directional and constantly interrupted (for example, by blinking), sound is omnidirectional, creating a context in which focus requires a particular kind of attentive filtering rather than a reframing or panning directionality. As will be taken up momentarily, listening and hearing are questions of filtering and attention rather than framing and focus.

Third, sound is and always has been. Mediated, manipulated, lost and found. Moreover, there is a deep history of relationships among science, medicine, philosophy, and the aural that is often overlooked (Erlmann, 2010). Reason is not necessarily separate from resonance nor, as scholars in sound studies document, is the sonic apolitical, ahistorical, or otherwise distanced from questions of positionality and power (e.g., Bull & Back, 2003; Sterne, 2012). Sound then can be understood as a strong means for philosophizing—an aspect of experience that has deep ties to the beingknowingdoing, from premodern to the post-nexts, that can offer pathways to queering the pitch (Brett, Wood, & Thomas, 2006) of everyday understandings.

Vibrational affect is at once central to collapsing relationships between humans, ecologies, and things, while retaining the significance of humans, their impact on one another, and to questions of power and the political (Gershon, 2013, Daza & Gershon, 2015). Rather than return to constructions of framing and sight, this chapter takes on sonic possibilities for philosophy in general and for educational philosophy in specific. Unlike sight and sighted metaphors, the most directive and closed of possible avenues in/with/through the sensorium, sound philosophizing, along with most other always already engaged senses, requires attention to filtering rather than the kinds of focus that are central to sight and ocular metaphors. Understandings are kinds of resonances, movements across/through barriers both metaphoric and practical. If all things vibrate, then all things have the potential to vibrate with all other things, both internally and externally, over history and time, space, and place. Yet, sounds are never pristine, bouncing a bit on this and that, and are always to some degree misheard. There are reverberations and echoes of that which once was, from the theoretical to the haptic, from the sound of your mother's voice to gravitational waves.

The ideas presented in this chapter will undoubtedly seem familiar to those with knowledge of the works of such philosophers as Walter Benjamin, John Dewey and Arthur Bently, William James, Charles Pierce, Baruch Spinoza, and Alfred Whitehead. It will be similarly familiar to those with an interest in non-Western spiritualities and philosophies, understandings that have no need for postmodern, poststructuralist, or current movements of the "post-next," for there never was the error of constructing false binaries between objects, ideas, ecologies, and the like. For example, there is no separation between people, ideas, ideals, things, and ecologies in many forms of Buddhism, an understanding that seriously questions not only Western notions of the five senses but also the continuing split between ontology and epistemology, even when noted as "onto-epistemic."

Where this work breaks new ground is in enunciating what it might mean to construct an educational philosophy around the ear, sound, and music rather than the eye, sight, and visual metaphors. In keeping with the interdisciplinarity of sound philosophizing, this chapter draws from multiple fields, each of which are in turn also interdisciplinary: sound studies, sensory studies, affect theory, and curriculum studies, fields that can be understood to reside at the complex nexus of philosophy, social theory, discourse, and materiality. Of primary importance are recent discussions about sound, vibration, and resonance, and it is these works that serve as the foundation for most of the arguments made here. These understandings are then displaced against interpretations in the social and physical sciences to form an educational philosophy of resonance, possibilities that also speak to questions about contemporary educational policy and practice.[2]

Here, I seek to more deeply explore these proposed possibilities and, in so doing, attend to what they may yet mean for educational understandings, particularly in the field of curriculum studies and associated fields such as educational foundations, philosophy of education, teacher education, education for social justice, and urban education. It is a point of wonder: What might attention to education through an otology of the sonic imply and engender for educational understandings?

My approach to this argument is through four overarching moves. In the first, I attend to vibrational affect and its implications for conceptualizing ideas and ideals as forms of resonance. Next, I articulate possibilities for resonance and its expression as reverberation, the inherent metaphoric and literal movements of resonances. Such an articulation of resonance and reverberation presses at current conceptualizations of scale, the third move before arguing for the importance and centrality of deep listening to a sound philosophy of education. Along the way, I pick up arguments made by Anna Lowenhaupt Tsing, particularly building off of what she calls nonscalability (2012, 2015) towards points that speak to notions of sound and scale.

Vibrational Affect: Understandings at the Intersection of Physics, Social Science, and Sound

Drawing from James's notion of "pulsed vectors of feeling" and Spinoza's discussion of movements as well as "the potential of entities to affect and be affected," Steve Goodman (2010) argues for a "vibrational ontology," an understanding that "everything is in motion, is vibrating" (p. 83). To this idea, he adds the Spinozan notion that "a body is, not because it thinks, but because of its power to affect and be affected. And for Spinoza [and more contemporarily Maxine Greene], we do not yet know this power!" (Goodman, p. 101). When placed alongside one another, these create a context in which not only the material but also the philosophical and theoretical are vibrating in ways that are constantly and consistently affecting and being affected, and affect everything from ideas to ecologies. This construction echoes not only recent work in anthropology, such as Anna Lowenhaupt Tsing's (2005) notion of the generative nature, for better and worse, of points of friction but also Melissa Greigg and Gergory Seigworth's (2010) discussion of affect as emergent "bloom spaces" of possibility (p. 12). Or, as Anahid Kassabian (2013) argues in *Ubiquitous Listening: Affect, Attention, and Distributed Subjectivity*:

> Affect is the circuit of bodily responses to stimuli that take place before conscious apprehension. Once apprehended, the responses pass into thoughts and feelings, though they always leave behind a residue. This residue accretes in our bodies, becoming the stuff of future affective responses.
>
> (p. xii)

That everything is in motion is also an axiom that lies at the core of much of contemporary physics—nothing is still. The same can be said metaphorically of ideas. There is no stillness to thoughts; even repetition requires motion as does the effort to keep believing the same thing over time, like a kayaker pressing back to stay in place in spite of a never-ending flow of possible paths.

Such motion has similarly been visualized at the intersection of complexity science and mapping, noted for its queerly somatic possibilities, and the ordinariness of its iteratively recursive nested layers in everyday experience has been sensuously documented. In short, from physics to social science, from the theoretical to the material, there is an understanding in many corners of contemporary thought and practice that all things vibrate. Of equal importance, these vibrations are what Brian Massumi (2002) calls "ontogenetic" (p. 8), a term that I have adopted with a slight change as ontogenic, a move meant to further emphasize the continual state of ongoing is-ness that is the movement of the affective.

Sound is vibrational, and the limits for its perception are embodied limitations of sensation. In this way, not only does sound interrupt many commonly held philosophical constructions, such as the discussion around the *augenblick*, for there is no blink of an ear (Kim-Cohen), but it also disrupts similarly common understandings about both perception and embodiment. There are vibrations that sound both below and above the human range of perception, the aural, the somatic, the intuitive, and other aspects of the sensorium (Goodman).

Vibration, like sound, "radiates spherically from a sounding source" (Price, 2011, p. 20), a material manifestation that similarly calls into question philosophical understandings that are based on ocular metaphors and visual meanings. For example, as was further brought to bear in the previous two chapters and will be again taken up in the methodology section of this book, the spatial limits for educational transactions are often more aural than visual. These spheres of vibration are at once literal and metaphorical, like an omnidirectional microphone, simultaneously vibrating and vibrated by the nested ecologies in which they exist. While this understanding of fields is in many ways similar to Pierre Bourdieu's (2012) construction of the same name—the idea that people are affected by and affect multiple layers of ideas, power, and culture—it is closer to Sherry Ortner's discussion of agency:

> The way I envisage social agents, which is that they are always involved in, and can never act outside of, the multiplicity of social relations in which they are enmeshed. Thus while all social actors are assumed to "have" agency, the idea of actors as always being engaged with others in the play of serious games is meant to make it virtually impossible to imagine that the agent is free, or is an unfettered individual.
>
> (p. 130)

On this multiplicity of social relations, Ortner's (2007) construction of agency can take one of two central forms: "enmeshed in relations of (would-be) solidarity" such as "friends, family, kin, spouses or partners . . . and so forth" or "enmeshed within relations of power, inequality, and competition" (pp. 130–131). It is an understanding of relations and relatedness that jives with other discussions about connectivity, the impossibility of the singular, and the production of movements of all kinds in contemporary social theory (e.g., Spence, 2015; Taylor, 2016; Tsing, 2005).

Similarly, as with aspects of John Dewey's (and Arthur Bentley's) discussion of transaction (1949), and contemporary notions of physics, as well as longstanding understandings in Buddhism(s) and a multitude of Indigenous ways of beingknowingdoing, if everything vibrates then nothing is unconnected or can be successfully decoupled from another. The difficulty with transaction, unlike Tsing's (2005) self-proclaimed poetically ethnographic study of destruction, is that there is largely not a politics (read: power and relations) of transaction. In addition, although both transaction and friction are certainly affective and sensual, both are often grounded in the ocular, and transaction more so than friction.

What construct, then, might encompass the wide range of possibilities of an ontology of vibrational affect? An answer, and the one that I propose here, and the one around which this paper is constructed, is the same that is used in sound studies, social theory, philosophy, social science, and the sciences: resonance.

On Resonance: Possibilities Beyond Consonance, Dissonance, and Relevance

If everything vibrates, then everything resonates. Resonance is produced by the oscillation of vibration, the peaks and valleys of something in and out of phase with itself and its surrounding nested layers of ecology. According to this understanding of resonance, everything has an iterative and recursive relationship at some level to itself and things not-itself. Rephrased slightly, everything resonates with itself and other kinds of things in ways that, while not predictive, are usually patterned—patterns that are in turn repeated so that they might be recognizable in kind. In this way, resonances can help identify possibilities of pattern while avoiding deterministic assumptions of them. Resonance therefore simultaneously denotes a particular kind of internal relations and the relationship between internal and external relations.

Musically speaking, resonances are often presented in relation to the constructions of consonance and dissonance. Rather than walk down this road of how a sound relationship is defined as consonant or resonant, particularly as is often done utilizing Western art music[3] and its relationship to pure harmonies first ascribed by the church in the Middle Ages, I wish here to note that consonance and dissonance are sociocultural constructions. What is dissonant in one set of musical understandings can be consonant in another, the way a flat seventh can be dissonant in Western art music and consonant in the blues and many forms of jazz for example. To push at this point further, even jazz, with its similarly limited set of 12 possible tones, as compared to the so-called "microtones" of Carnatic (South Indian) or *musqi-e assil* (Iranian) "classical" music for example, often has tones that one is to play or not play in order to be consonant or dissonant with a given chord structure. Where resonance is inevitable, consonance and dissonance are constructed.

If consonance and dissonance are sociocultural constructs, then the particularities of their categorization are even less significant, at least for the argument here, in comparison to their presence as forms of resonance. In traditional Western music theory, consonance is most often applied to sounds/pitches that cause sympathetic resonance, to produce an overtone series or when playing a particular note/sound/pitch causes another object in a room to audibly hum for example. Dissonance, in turn, is those sounds that do not produce such reactions. These distinctions, however, quickly fall apart when volume and loudness are applied, making even supposedly dissonant objects noticeably vibrate. As described above, what is dissonant to one group is consonant to another, then such distinctions are contextually significant but similar in kind. What is up to debate is how given ideas, actions, norms, values, ecologies, possibilities, etc., are apprehended but *not* either their affect or their impact.

This creates the conditions in which conformity to how things should be/sound is consonance with a dominant sound. Nonconformity is dissonance when compared to the dominant sound but is nonetheless resonant. At the same time, dissonance is necessarily consonant with another sound. The greater the dissonance, the wider the spectrum of possible congruent resonances, the greater the opportunity for iterations of consonance.

Regardless of their perceived consonance or dissonance, all sounds are resonant.

Philosophically speaking, affects and thoughts resonate. They resonate within and between systems: embodied/material, theoretical/discursive, self/other, object/ecology, feeling/thought, and so on. The ontogenic and epistemogenic are similarly processes of resonance. History and culture are resonant as are ordinary events and experiences. What is sensible makes sense, its processes of sensation and signification, through resonance. Resonance then bridges the gap in human experience between unique individuality and the sociocultural contexts that inform that individualism—resonance must exist both within and outside of the self.

Of perhaps even greater importance, sounds make clear continuing not-so-implicit false binaries between ontology and epistemology all the more evident. What has come to be called "the ontological turn," in some circles, continues to strongly inform a wide swath of scholarship across the social sciences and humanities. What is often lost in such understandings, or is inverted with ways of knowing placed within ways of being, is that a reframing of the ontogenic as primary nonetheless maintains an ontological/epistemological binary. Even as Sam Rocha (2014) enunciates the differences in Spanish between types of knowing—*conocer* as knowing another's "heart" and *saber* as knowing information—sounds do something else entirely: they note the always intertwined nature of beingknowingdoing or, as Bob Marley sang: One good thing about music when it hits you, you feel no pain. The sonic is always vibrational affect that carries information that does something to you—from caressing the cilia in your inner ears to the hairs on the nape of your neck, that is, in turn, the aftershock of something else—as is explicitly clear in both the recent success of hearing gravitational waves and in Jenna Levin's (2016) recounting of the history of this project in *Black Hole Blues and Other Songs from Outer Space*. All at once. All of the time. Beingknowingdoing.

Of equal importance, in the beginning of her book during a hypothetical exercise conceptualizing what it might be like for an astronaut in space at the moment of a black hole collision Levin informs readers that, "an astronaut floating nearby would see nothing. . . . If close enough her auditory mechanism could vibrate in response" (p. 3). Levin continues, "gravitational waves are like sounds without a material medium. When black holes collide, they make a sound" (p. 5). An understanding that one can hear something without a material medium at once supports a metaphorical notion of the possibilities for hearing ideas and ideals resonating across literal and metaphorical spacetime (e.g., Collier, 2014; Taylor & Wheeler, 1992) while pressing at the notion of sounding. Taken up in a particular fashion, and one I'm inclined towards here, the notion of being affected by unheard resonances speaks loudly to conceptualizations of scale, immediacy, and impact, points to which I return below.

This raises an important question: What isn't resonance? One answer to this question is also found in its construction. Simply because something resonates does not mean that a given idea, object, ecology, process, or ideal is resonant with everything, most things, or anything at all. For example, the work of a social scientist can be conceptualized as an effort to better understand how actors' sense-making resonates with other actors as well as with nested layers of sociocultural norms, values, systems, and processes (e.g., Paerregaard, 2002, Young et al., 2010).

Additionally, if everything resonates, then one must attend to how such resonances might be resonant *within* any person and *between* any person and any other object, idea, ideal, feeling, process, or ecology. If resonance occurs within and between this endless array of possibilities, then each person should likely be trusted to know, at some level, their own sense of resonance and what is resonant to herself. Because of this likelihood, each person should therefore also be given the dignity of such positionality.

This is not to say that resonance is without politics, position, or power, quite the opposite. A given resonance can hold any variety of embedded norms and values, explicit or implied. How they are received and taken up is therefore an indicator of not only the kinds of possible resonances available but also the kinds of resonances to which one is attuned. It is by granting a person the trust and dignity to enunciate in some fashion (intention, action, verbalization, and so on) what they find to be resonant that a given resonance can be strongly critiqued. For only when a resonance is respected can it be fully engaged.

As but one of any number of possible examples, in this way, rather than dismiss an act of hate speech as unintentional, naïve, or unaware, taking the possibility that certain racist ideas and ideals resonate with the speaker and giving the speaker the dignity of having those ideas provides an ethical foundation for strong critique. Even if that idea *can* be primarily attributed to a lack of experience or understanding, the speaker, at that moment, resonates with particular racist views. Resonance is therefore necessarily political, an alignment between self and world according to particular sets of processes, feelings, ideas, and ideals. Rather than removing either power or position, resonance instead emphasizes both.

Given this construction, resonance has particular advantages over relevance as a means for considering ideas and is equally valuable for the consideration of affect and process. Relevance is the consensus process through which various kinds of resonance are deemed not to be of value or where particular kinds of resonance are adjudicated and rendered valuable. As Jacques Rancière (2010) has argued, processes of consensus tend to form limiting, closed systems through which particular people, groups, ideas, ideals, and processes become ethical at the expense of others' values, processes, ideas, and ideals. This is in no small part because consensus is inexorably linked to conformity and conformity is a key mode of oppression.

Where relevance is a consensus perspective that seeks to understand the world by valuing some ideas at the expense of others, resonance accepts that all ideas are possible and provides a space for critique. Where possibilities of and for resonance are more practically applied to education, deploying the category of relevance is a possibly dangerous and damaging act. For example, there is a long history in schooling in the United States of marginalizing people of color through discussions of relevance (Cooper, 1892; Rosiek & Kinslow, 2015; Winfield, 2007; Woodson, 1933).

Yet, resonance is always the result of waves in motion. It is therefore always in some fashion twinned with reverberation, the distribution and redistribution of those waves over not only things, ideas, ideals, and ecologies but also over spacetime. They are resonant because of reverberations.

On Reverberations

Reverberations, like resonances, are physical and metaphorical, sensorial and signified. Because sounds are always in motion, they are always reverberating, bouncing off of objects from sinus cavities to walls to coral to brush to air to water to stone. The always-mediated nature of the sonic pushes the nonsensical to the fore, an acknowledgement that the sound at the point of perception is not what it once was but as it is received. Even sounds generated by the body receiving them are necessarily mediated, not as they were but as they are heard. Mishearing is the norm, not the exception.

Unlike sight, then, with its assumptions of linearity and side-by-side sequencing, sound is a mishmash of moving, messy, omnidirectional, multisensory, polyphonic, mediated, aurality, bouncing from and through things in ways that sight does not. To say that sound blurs the lines between ecologies, things, layers, events, and experiences is in error because for sound those boundaries do not necessarily exist in the first place.

Although such understandings speak directly to interpretations integral to (qualitative) research methodologies as is taken up in Chapter 5, they also enunciate clear philosophical issues with taken-for-granted foundational perspectives in education. For example, educators often speak of necessary foundations for building knowledge, metaphors of architecture that are, in turn, sighted. What happens to education philosophically if its central metaphors, for an education both as it is desired and dismissed, not only no longer hold water but also cannot possibly be imagined to hold anything in its grasp? What happens when foundations become resonances and reverberations?

To begin, sounds are at once physical, liminal, omnidirectional, polyphonic, and mediated, vibrating that which they impact and impacted by that which they vibrate. Echo chambers, of either the metaphorical or material variety, not only amplify the information they receive but also further distort those messages on at least two levels: 1) the continual attack and decay of the waves as they bounce within and around the chamber and 2) the increasingly inaudible content of overlapping messages that often decrease in sensibility as they increase in amplitude. Further, the walls of the echo chamber can be more resonant than the mediated sounds they contain as they crash into one another within its confines. Yet, we focus more on that which happens within its confines than that which escapes, information that is a constant flow out of the chamber as much as within it. The reason for its success is similarly sonic, sound as misdirection, drawing attention through consistent volume, tone, and pitch, aided by continually pressing sound/information into the echo chamber, snapping our metaphoric heads in the direction of that which is most audible and away from the equally present flow that is constantly escaping its boundaries.

Echoes coat the stillness, eddies of sound possibilities bouncing about, underscoring the impossibility of a sterile sound environment. Water is an inept metaphor for, as much as water seeps through cracks and crevices, sound travels through walls, soil, liquids and solids and gases. Not that sound is an immutable all powerful motion but rather that it is an impossibility to fully stop or completely reduce the ever-occurring polyphony of vibrational affects.

Sound and Scale

Tsing's (2012) notion of nonscalability is an example of what is at once so very helpful and possibly concerning about scale, underscoring how notions of scale are inherently imbued with sight and, as such, necessarily fall flat, both literally and metaphorically. For Tsing, there are two concerns with scale, the first of which is that scale is at once a tool for ever-enlarging processes so that interactions remain constant while they grow in size. Scale becomes a tool for "scaling up," moving procedures and practices that are locally useful in their ability to effectively ensure reproduction that enhances monetary gain to an ever-widening group of workers.

Groups and the individuals that comprise them work hard to create processes that can become "scalable" or, as Tsing argues, "Somehow, project elements had to be stabilized so that expansion added more elements without changing the program" (p. 507). Scalability is therefore socially and intentionally constructed. Also embedded in these understandings is the use of scale and its promotion by economic interests in order to increase profit, and an accompanying understanding that the mobility of moving projects across ever greater layers of scale are sets of competencies through which culture is mobilized (Tsing, 2015, p. 509). This is because it is through such mobilization that scaling up is not only an increase in reach and expansion of potential profits and associated labor needs and practices, but is also a conduit for particular sets of sociocultural understandings.

In these ways, scalability is not only a normalized global enforcement of particular norms and values, those that help those doing the globalizing get further reach and increase their income, but also the patterned privileges that enable and engender possibilities for expansion. In sum, Tsing argues that scalability is one of the central, if not the central, mechanisms for capitalization in action, a feedback loop in which expansion is economic advancement, and advancement is, in turn, the further mobilization and application of the scalable.

Second, the artificial normalcy of scalability creates what Tsing calls "nonscalable," the endless things that cannot be readily scaled, from ecosystems and ecologies to individual things and their interactions, regardless of their complexity. Nonscalability is an examination of those things that cannot be scaled and the detritus and destruction that scalability leaves in its wake once it can no longer sustain either its current status or further iterations of expansion. "Nonscalability theory is an analytic apparatus that helps us notice nonscalable phenomena" (p. 509). It is by attending to nonscalability that one can both see and address scalability and its contingent concerns. To further drive this point home, Tsing notes that both Walmart and Amazon have successfully utilized several key facets of nonscalability including independent contracting and the distribution of goods at low prices with far less attention to the means and processes that allow products to be passed along at such rock bottom prices.

This is because, as Tsing (2012) makes clear, nonscalability has neither a necessary morality nor an inherent ethic. Nonscability theory is, however, a strong means for attending to both the multitude of narratives outside of scalability narratives, including Marxism (p. 522), and how the varieties of such phenomena/experiences/interactions deeply inform what is scalable and the ways in which those processes unfold.[4] Tsing's theorizing here is incredibly helpful if one continues a sighted consideration of scale, which, although it can certainly be applied to sound, is most often conceptualized as an ocular construction and metaphor.

From a sonic positionality, scale has neither consistent boundaries nor cleanly demarcated borders. Sounds press through the physical, cut across time, and are simultaneously deeply embedded with intended and perceived information and liminal, only reproduced through a modern trick of the tale, a construction that claims recorded sounds are real and accurate reproductions. Enunciating the constructed nature of recording is important here because recording processes are in many ways based on rhetorical corrections that hide the inherent messiness of the practical: spaces in which sounds are recorded overlook the impossibility of truly accurate representation, the sculpting of sound through machines is passed on as largely unmediated sonic reproduction, and their storage and dissemination to others is conceptualized as accuracy. It is a set of understandings that allows a person who was at a live recording to consider a recording of that auditory experience as, in larger part, "the same" as their experience and that of others at the same event, in spite of the fact that such an experience is impossible from personal, philosophical, and scientific perspectives. It is a trick of interpretational will.

Interpretational will is of no small significance. We can recognize a person's voice over the phone in spite of the paucity of sonic clues and clarity. Metaphorically, what is important here is the recognition that hearing is as much an inevitability as it is constant acts of interpretation. How one is attuned, to what one attends, and that which sings one's tunes are all in play. The combination of soundwaves reverberating across spacetime, the cacophony of multiplicities in their varied simultaneity, and their inability to stay firmly in their level of scale create a context in which, although sounds occur at varying distances from a person, thing, or ecology, can be apprehended by an individual regardless of their distance from those events.

Such attention does funny things with scale. Not so much a blurring, for that is too static, too clear, too ocular. It is more akin to the inherent messiness of waves crashing on you, on distant shores, as they stir yet other ripples that cannot help but to further amplify that which cannot ever be distilled but necessarily roll over you, puddling at your feet until you realize that this is literally and metaphorically but a drop in the ocean of what is, was, will, might, should, or could ever be. And so, scale is real and unreal, stands strong and collapses. For it is certainly the case that that which happens to you is to some degree removed from that which happened less close by, that which are again farther away until they operate so far afield in ways that press at your capacity to imagine, like a different plane of existence on another galaxy in an alternate universe.

More concretely, a sonic perspective of scale creates a context in which vibrations, resonances, and reverberations echo from far away and then immediately nearby, within you for example (e.g., Gandy & Nilsen, 2014; Gershon, 2011b; Goodman, 2010). Together, they create a continuous polyphonic echoing of reverberating resonances. It is also why individual, institutional, and systemic injustices, although occurring at increasing levels of scale away from one's self, are utterly collapsed in a single iteration of interpersonal aggressions and how interpersonal aggressions combine to create systems of injustice across scales—a feedback loop of oppression, marginalization, and aggression (e.g., Browne, 2015; McKittrick, 2006; Shabazz, 2015). According to this articulation of sound social theory, there can be no microaggressions because each aggression carries the resonances across/through all other layers of scale. Fortunately, the same is true for love.

Speaking to the points Tsing (2012, 2015) has so elegantly raised in her recent works, from this messy, never-ending set of sound possibilities, there is no possible separation between scalable and nonscalable. For they are, as she notes, part and parcel of the same sets of resonances, one heightened, another other dampened. Conceptualizing scale through sound creates a means through which the complexity of distance from a starting point is maintained but the experience of such imbricated messiness across/between/through scales is collapsed. This set of possibilities retains understandings of place as the throwntogetherness of history, geography, interactions, processes, and ecologies as noted in the previous chapter (Massey, 2005)[5] as well as its liminality. It also presses as such constructions of place in that sound, although certainly dampened by many materials, is by no means stopped by them, while asserting the necessity of contributions from things, people, animals, ecologies, and even the continual construction of place and space back outwards towards the inputs received. Always a feedback loop of liminal vibrations that follow and lead and express and depress.

Finally, sonic enunciations of scale call for a different articulation of space and place, one that manages to acknowledge notions of physical distance and material things—just because sound readily moves through physical things doesn't mean that those things don't exist—while making the fluid nature of space and place more audible for the listener. One such possibility is a reconsideration of fields. Fields have a long history of use in the social sciences and are at once specific enough to encircle a specific area yet open enough a descriptor to attend to the fluidity, porousness, and liminality of sound. Fields also can be shaped in any fashion, considered quantitatively, qualitatively, and through a wide variety of media, and can speak at once to a multitude of constructs (e.g., race, class, gender, space, place), modes of expression, and degrees of reception.

For example, as presented in the previous chapter, a street in front of a school and a school are often part of the same sound field even though they are often described as distinct physical spaces and layers of scale from an optical perspective: classroom, hallway, school, play yard, street, neighborhood. Yet, sounds from the street impact the sound in a classroom and vice versa; the sound of a school bell regulates the time not only for those in school but also for the surrounding neighborhood. Fields can have density, depth, tenor, and tone, so that a given field can be presented as either more singular yet polyphonic or an overlapping multiplicity depending on questions of context, intention, argument, and the like. Further, physical places and spaces are not erased by this notion nor are theoretical possibilities. Sounds are enhanced and dampened by material contours from bodies to ecologies, decay over time and distance, and are subject to both questions of quality and measurement. From this orientation, sound fields at once maintain the essential features of scale—distance, time, space, place, ecologies, and the things that populate them for example—while adding to our understandings in ways that further complicate and make complex the inexorably polyphonic notions at the forefront of contemporary understandings in critical and human geographies (e.g., Massey, 2005; McKittrick, 2006; Shabazz, 2015). Such an understanding is also commiserate with indigenous discussions of space and place (e.g., Kermoal & Altamirano-Jimenez, 2015; Sayers, 2016) and ongoing discussions of space and place in curriculum studies (e.g., Helfenbein, 2006, 2009; Ng-A-Fook, 2007; Tuck & McKenzie, 2015; Whitlock, 2007).

In short, we move in a constant form of theoretical and material echolocation. Not solely inasthrough sound for we are not only the aurally haptic sense(s) we call sound—we are also not simply sighted, yet our metaphors and material understandings are crushingly dominated by the ocular as noted in the previous chapters and again above. It is therefore no less a possibility to enunciate (rather than envision) a sonic reorganization of possibility for the consideration of educational theories, methods, and practices.

Deep Listening: Sound Educational Philosophy

At the heart of anything that might be called a sound philosophy of education or a sound educational philosophy lies a kind of deep listening (Oliveros,

2005) in which those who are giving voice believe themselves to be heard. There is a significant difference between a more pedantic construction of providing others the space to be heard—a move that retains the overarching power of and discretion to give voice—getting credit for telling others' stories, and a practice of deep, intentional listening in ways that provide a critical space for expression, especially for those who are often silenced or whose perspectives reject, resist, or otherwise interrupt dominant, normalized discourses and the values they present.

Deep listening provides an opportunity for often-dampened resonances to be heard, to allow for the inevitable mishearing of reverberations without a need to correct or silence, and for an understanding that what is voiced is of significance to the speaker. It begins with an act of intentional attention to the expression of others. As but one possible concrete example, of the over 20 interviews with undergraduates of color about their experiences on campus, a space my colleague Rob Helfenbein calls a "ridiculously white campus," the overwhelming answer to what they would like from their rather Anglo faculty is for faculty to listen and take students' perspectives seriously. This, they affirmed, is the single biggest difference between faculty of color on campus, even those with whose politics and practices they disagree, and Anglo faculty: their time and ear regardless of the consonance of their perspectives.

What might a sound educational philosophy contribute to our current conceptualizations of education as well as for possibilities and challenges for educational beingknowingdoing? This is a tricky pivot for sound philosophizing as it requires landing intentionally open and emerging developments in sticky (Ahmed, 2010) material matters. For this reason, I wish to pause for a moment to reiterate that I understand the potential work described here in general and in this section more specifically as but one set of possibilities at the intersection of sound, philosophy, and education. As such, the examples provided here are precisely that, selected pathways among many options brought to bear as a means to articulate its potential practicality.

To begin, a sound philosophy of education erases long-held distinctions between epistemology, ontology, and enactment. Not that there aren't ways of thinking, being, and doing but rather that such distinctions are neither clear-cut nor as independent as they are often conceptualized across educational literatures. Once collapsed as beingknowingdoing, approaches to teaching and studenting—performing the sociocultural role of student—can no longer be falsely split into long-debunked Cartesian splits of mind/body that underpin contemporary educational matters from lesson planning to discrete measurable skills. Beingknowingdoing also calls attention to how sociocultural contexts are necessarily part and parcel of everyday classroom life and the impossibility of separating anything that might be conceptualized as academic from other things that could be called social or cultural.

Philosophically speaking, beingknowingdoing causes some potential issues and disjuncture for philosophies of education that have a long history of attending to being, knowing, and doing as separate and not necessarily related activities/experiences. However, education, at its most basic level, is necessarily an action (doing) in which how one is (being) deeply influences how and what one knows (knowing). A student's experience and feelings about tests and taking assessments impacts what and how she knows information presented on that test and how she engages the action of testing.

That all things are resonant and have the potential to resonate with anyone presses at questions of curriculum, instruction, and practice. Education has long been a question of relevance: What ideas are most worth knowing? Often missing from this construction are questions about to whom such knowledge is valuable, in whose interests is that knowledge reproduced, and the social messages present in a given set of curricular information. Sound educational philosophy insists that such possibilities be taken into account.

That people receive resonances as reverberations speaks to the impossibility of correctness, for all possible perceptions are necessarily misheard. Although this might seem to be a distinct concern for education, it is in fact a boon. For it leads yet again to a multiplicity of potential pathways of understandings while insisting that any singularity, although helpful for some, is but one possibility among many and, as such, creates opportunities for grasping any given idea from a wide variety of trajectories.

Resonance and reverb also refocus educational theory and practice on questions of ethics and inquiry. Having an infinite number of conduits and approaches to a given event, situation, or question means that one is likely to have to justify why any one particular choice is suited to that event, situation, or question. Such discussions are also much more likely to engage neighboring concerns and possibilities so that a choice is not merely a procedure for its own sake (getting a correct answer) but is instead a directionality related to contextualizing factors. It is, in other words, a pathway of inquiry according to a particular set of understandings.

Attending to a rationale behind a decision, what is in many ways the kinds of attention educators intend when they refer to something as thinking within a field (e.g., historical thinking), also lays bare an attending set of ethics that govern that pathway to inquiry. Not that a sound educational philosophy is necessarily ethical per se, as an ethics is not in and of itself the kind of conscientious, socially just path that educators often believe lies at the center of their practice.

For example, Franklin Bobbit, in many ways the founding father of educational curriculum through his 1918 work *The Curriculum*, was a generally nasty piece of work who hoped to engender his eugenic ideas and ideals through educational practice. His touting of measureable goals and objectives, later refined by Ralph Tyler (1949), and the kinds of attention to everyday practice is still how we teach today. Ends-means educational practice regardless of content.

Collapsing scale means that the continuing reverberations of measurable goals and objectives, despite deep critique (e.g., Kliebard, 1975; Woodson,

1933) and alternative possibilities (e.g., Doll, 1993; hooks, 1994), still contain long histories of eugenics, othering along racial, class, and gender lines, and the masking of those tendencies through assessments that use their surface objectivity to mask their deep histories of intentional marginalization by race, class, and gender (among other similar baggage). That I can cite counterexamples is another manifestation of how resonances and reverberations function, dampened but not silenced narratives echoing over spacetime.

It is important to note here that this is a particular kind of interruption, one for a polyphony of Western, Eurocentric educational practices that are manifest as contemporary schooling in the United States. Although this argument for a sound educational philosophy would certainly affirm or disturb (or both) other "Western" notions of education, depending on the kind of education, nation/state, and the like, it may not do so for other educational traditions. Much in the same way that I have argued that new materialisms is a strong means for interrupting Western-oriented social and political theories (Gershon, 2016), sound educational philosophy is not necessarily news to those who have not overly adopted a generally Westernized set of educational practices.

However, and herein lies the rub, most of my educational experiences have largely been through teacher-centric, top-down, ends-means educational practices. With the exception of three years at a private progressive elementary school and time teaching critical and creative thought to gifted and talented youth, a point to which I will return momentarily, this has been my life and tendencies that I fight as an educator. What makes this striking is that I have been educated in a wide variety of traditions, from public education in the U.S. to hundreds of years old oral traditions, and most have been some version of the kinds of education most kids receive in U.S. public schools.

On one hand, there is good reason for this kind of educational practice. Teachers are supposed to have knowledge that students do not, to impart that knowledge in a clear, thoughtful, and respectful fashion, and to attend to students as they come to better grasp those understandings. Or, as Maxine Greene stated, "I don't think any of you signed up for this class to hear from each person equally. I certainly want access to your needs and desires, but I am not enslaved by them. I want to welcome your responsiveness, too, but let's not make a fetish of chairs in a circle" (Ayers and Miller, 1997, p. 8). Before moving too quickly to a discussion of this as a privileged perspective, it is important to note that those of us who regularly teach topics in required courses relating to sociocultural constructs like race, class, gender/identity/expression, and sexual orientation to majority Anglo, suburban, middle class students also must insist on our perspectives and/or keeping to particular educational visions in order to interrupt rhetoric and material practices that (un)intentionally cause harm to traditionally marginalized populations.

On the other hand, such practices make false assumptions about students, their experiences and understandings. If you already know what you're going to teach before you begin, move on to the next lesson regardless of students' comprehension, and assess them on tests made on what students should know rather than what they're taught, what is the purpose of having students or for students to engage in educational interactions?

A move from sight to sound is also a move from processes framing to processes of filtering. Unlike eyes that frame and can blink, we are constantly bombarded by and emit sounds regardless of our ability to hear them. This is not solely or even primarily a question of being deaf or hearing. Sub- and hyper-sonic sounds nonetheless affect our lives, and our ability to perceive them shapes our metaphors, the ways in which we conceptualize ecologies, relations, and things. This difference can be understood as the distinction between finding a person or listening to a conversation in a crowd—where the former is tacking back and forth through visible frames, the latter is a constant question of attention and inattention to a constant flow of simultaneous input. Philosophically, this brings us back to questions of beingknowingdoing and the intentionality of filtering.

Again, resonance is not apolitical or all powerful. Some ideas resonate more strongly than others and their presence requires constant maintenance. Yet, everyone does not have the ability to maintain a given way of being or knowing equally. However, as social scientists have strongly argued, a lack of power is not the same as a lack of agency, and the inability to overly impact one's context is not the same as an inability to feel or think deeply in ways that can resist, reject, reify, or otherwise engage both self and context(s). Echoing a point raised in the previous section, if all ways of being, knowing, ideas, ideals, or process might resonate with any given person, then each person should be given the trust and dignity of that resonance.

Aesthetically political and politically aesthetic, ontogenic and epistemogenic, a sound educational philosophy emphasizes responsibility over accountability, flux over stability, open over closed systems of knowledge, and inquiry over known-information questions. It suggests, somewhat paradoxically, that the greater the dissonance in a learning ecology the greater the opportunity for resonance; the more room for diversity (of feeling, thought, process), the greater the likelihood that a student will resonate with a given idea, ideal, or process. It also speaks to the importance of intentionality and intuition for all educational actors, while providing the means for strong critique.

Afforded the trust and dignity of internal and external resonance, each dependent on the other, creating a context in which students must be respected, encouraged, and critiqued; where teachers should be given the space to educate as they understand their work to positively impact students and a similar space to be critiqued by others (especially) including students; and where teaching and learning are emergent activities. Resonance also underscores the political actions of choice and the operation of power. If all ideas are possible, resonance provides an opportunity for multiple roads to success and understanding while emphasizing curricular and pedagogical choice, this idea over that, this teaching tool over that. A philosophy of educational resonance regards how knowledge, feelings, and processes are vibrational affects that are connected and interconnected in expected and unexpected ways, expectations that are, in turn, dependent upon a constantly fluctuating transaction of resonances and the reverberations they produce.

Coda: Sonic Everywheres, Everywhens

There are no geographical claims to sound philosophies of and for education. Nor is there a mandatory theoretical pathway for its conceptualization or expression. To be explicit, sounds are not the sole purview of the West, the global North, or philosophies of Europe and North America. As occasionally surfaced in this chapter, sound philosophies are endemic in all groups and individuals, each with their particular and equally significant understandings of sounds, relations, ecologies, and expressions. Sounds are of all things, across all ecologies, and over spacetimes, if even such things are not also falsely constructed as various Indigenous and other traditions suggest. It is difficult to argue for a right if everything is always already misheard.

Then there is the question of the specific, the wheres and whens of resonances, reverberations, and scales. Just like the media through which sounds travel, air, water, rocks, everything, as Lisa Gitelman (2008) suggests about media and Harve Varenne and Ray McDermott (1997) argue about sociocultural norms and values, sounds were always already there and new. They reverberate across everywheres from everywhens. The sonic was never not present nor somehow unavailable. They bounce along the millennia from every direction moving across scales and collapsed in an instance all at once, everywheres and everywhens.

Perhaps most importantly, a sound educational philosophy queers the pitch (Brett, Wood & Thomas, 2006) of sighted possibilities. As Brent Davis & Dennis Sumara (2006) remind us, linearity, correctness, and straightness are constructions, an example of how dominant norms and values at once produce constructed narrowness as commonplace and the ways in which metaphors impress regimentation on material bodies, as not-straight begets not-normal and queerness becomes othered. Along similar lines, just as Phillip Brett, Elizabeth Wood, and Gary C. Thomas note how LGBTQ voices queer expected paths in (Western) musicology, Rashad Shabazz (2015) argues the inverse, that Black sexuality lies outside of heteronormativity and is therefore necessarily queer.[6]

Bringing these three perspectives to bear on the sonic, sounds are impossible to straighten and cannot be correct (always misheard reverberations). When utilized, sounds interrupt the linearity of sight, and necessarily lie outside boundaries of both dominant modalities of discourse and metaphor and any constructions of heteronormativity or singularity of understandings. In sum, sound philosophies queer education.

Notes

1 Given its complex and significant history, some discussion of *clave* is warranted. At one level, *clave* is a rhythmic, two measure pattern that, unlike many Western musical traditions, creates not simply a sense of time, often expressed in Western notation as a time signature, but also strongly informs what is melodically and harmonically possible. On the other hand, *clave* is deeply rooted in West African religion, sociocultural norms and values, histories of pride, expression, artistry, creative nonconformity, oppression, resistance, sublimation, subversion, intentional syncretism, and other such combinations of space, place, history, ideas, ideals, and practices. *Clave* is both the underlying pulse that exists not only in the music of African diasporas but can be applied as a means for interpreting all music and a metaphor for possibility and ways of beingknowingdoing, musical and otherwise. In both cases, clave is the literal and metaphorical key to African diasporas that flowed through primarily Spanish and Portuguese colonialisms: in Spanish a key, as in an answer key to an assessment, is a *clave*.

2 As noted in the copyright section of this book, parts of this work have appeared in the publication "Vibrational Affect: Sound Theory and Method in Qualitative Research," and as part of the Annual Yearbook of the Philosophy of Education Society in a piece titled, "Affect, Trust, and Dignity: Ontological Possibilities and Material Consequences for a Philosophy of Educational Resonance." This latest version of my thinking on the topic is indebted to Gert Biesta's critique and conversation about the version presented at PES 2015 and Vanessa Andreotti for her thoughts on my approach. I am also grateful for thinking through these ideas with Sam Rocha, Reagan Mitchell, and Peter Appelbaum, all of whom have been overly generous with their critiques and support over time.

3 Western art music is the term musicologists and other associated fields of music use to describe what is often commonly referred to as "classical music."

4 To be abundantly clear, I think this is brilliant stuff and watching it morph from this iteration to talk in her most recent work (Tsing, 2015) is a joy. My comments here are therefore not a critique of Tsing's theorizing but on the limitations sighted metaphors and what sound considerations of scale might bring most conceptualizations of scale do not.

5 This construction is not only heavily indebted to Dorreen Massey's (2005) now well-known definition of place in her seminal work *For Space*, but also uses one of my favorite descriptions about theory in her talk about place, it's "throwntogetherness" (p. 149).

6 The complete point is worth quoting here in full: "In spite of earnest efforts, Black respectability could not produce the political gains it sought. It could not counter anti-Black racism through desexualizing blackness, in large part, because Black sexuality—whether respectable or queer—is always outside heteronormativity. In short, Black sexuality is queer because it's Black" (Shabazz, 2015, p. 125).

References

Ahmed, S. (2010). *The promise of happiness*. Durham, NC: Duke University Press.

Ayers, W., & Miller, J. L. (Eds.). (1997). *A light in dark times: Maxine Greene in the unfinished conversation*. New York: Teachers College Press.

Bourdieu, P. (2012). *The field of cultural production*. New York: Columbia University Press.

Brett, P., Wood, E., & Thomas, G. C. (Eds.). (2006). *Queering the pitch: The new gay and lesbian musicology* (2nd ed.). New York: Routledge.

Browne, S. (2015). *Dark matters: On the surveillance of Blackness*. Durham, NC: Duke University Press.

Bull, M., & Back, L. (2003). *The auditory culture reader*. Oxford: Berg Press.

Davis, B., & Sumara, D. (2006). Complexity and education: Inquiries into learning, teaching, and research. New York: Routledge.

Daza, S., & Gershon, W. S. (2015). Beyond ocular horizons: Sound, silence and sonification as inquiry. [Special Issue: 20th Anniversary]. *Qualitative Inquiry, 21*(7), 639–644.
Doll, W. E. Jr. (1993). *A post-modern perspective on curriculum*. New York: Teachers College Press.
Erlmann, V. (2010). *Reason and resonance: A history of modern aurality*. Cambridge, MA: Zone Books.
Gandy, M., & Nilsen, B. (Eds.). (2014). *The acoustic city*. Berlin: Jovis; Pap/Corn.
Gershon, W. S. (2011). Embodied knowledge: Sounds as educational systems. *Journal of Curriculum Theorizing, 27*(2), 66–81
Gershon, W. S. (2013). Vibrational affect: Sound theory and practice in qualitative research. *Cultural Studies ←→ Critical Methodologies, 13*(4), 257–262.
Gershon, W. S. (2016). The sound of silence: The material consequences of scholarship. In N. Snaza, S. E. Truman, D. Sonu & Z. Zowiska (Eds.), *Pedagogical matters: New materialisms and curriculum studies* (pp. 75–90). New York: Peter Lang.
Gitleman, L. (2008). *Always already new: Media, history, and the data of culture*. Cambridge, MA: MIT Press.
Goodman, S. (2010). *Sonic warfare: Sound, affect, and the ecology of fear*. Durham, NC: Duke University Press.
Greigg, M., & Seigworth, G. J. (2010). An inventory of shimmers. In M. Greigg & G. J. Seigworth (Eds.), *The affect theory reader* (pp. 1–17). Durham, NC: Duke University Press.
Helfenbein, R. J. (2006). Space, place, and identity in the teaching of history: Using critical geography to teach teachers in the American South. In A. Segall, E. Heilman, & C. Cherryomes, (Eds.), *Social studies—The next generation: Re-searching in the postmodern* (pp. 111–124). New York: Peter Lang.
hooks, b. (1994). *Teaching to transgress: Education as practice of freedom*. New York: Routledge.
Kassabian, A. (2013). *Ubiquitous listening: Affect, attention, and distributed subjectivity: Affect, attention, and distributed subjectivity*. Berkeley, CA: University of California Press.
Kermoal, N., & Altamirano-Jiménez, I. (2016). Living on the land: Indigenous women's understanding of place. Vancouver, B.C.: UBC Press.
Kim-Cohen, S. (2009). *In the blink of an ear: Towards a non-cochlear sonic arts*. New York: Continuum Press.
Kliebard, H. M. (1975). Critiques of Tyler. In W. F. Pinar (Ed.) *Curriculum theorizing: The reconceptualists* (pp. 70–81). Berkeley, CA: McCutchean.
Levin, J. (2016). Black hole blues and other songs from outer space. Boston: Beacon.
Massey, D. (2005). *For space*. Thousand Oaks, CA: Sage.
Massumi, B. (2002). *Parables for the virtual: Movement, affect, sensation*. Durham, NC: Duke University Press.
McKittrick, K. (2006). *Demonic grounds: Black women and the cartographies of struggle*. Minneapolis, MN: University of Minnesota Press.
Ng-A-Fook, N. (2007). *An indigenous curriculum of place: The United Houma Nation's continuous relationship with Louisiana's educational institutions*. New York: Peter Lang.
Ortner, S. (2007). *Anthropology and social theory*. Durham: Duke University Press.
Paerregaard, K. (2002). The resonance of fieldwork: Ethnographers, informants, and the creation of anthropological knowledge. *Social anthropology, 10*(3), 319–334.
Price, P. (2011). *Resonance: Philosophy for sonic arts*. New York: Atropos Press.
Rancière, J. (2010). *Dissensus: On politics and aesthetics*. New York: Bloomsbury.
Rocha, S. D. (2014). *A primer for philosophy and education*. Eugene, OR: Cascade Books.
Rosiek, J., & Kinslow, K. (2015). *Resegregation as curriculum: The meaning of the new racial segregation in U.S. public schools*. New York: Routledge.

Sayers, D. O. (2016). *A desolate place for a defiant people: The archaeology of Maroons, Indigenous Americans, and enslaved laborers in the Great Dismal Swamp*. Gainesville, FL: University of Florida Press.

Shabazz, R. (2015). *Spatializing Blackness: Architectures of confinement and Black masculinity in Chicago*. Urbana, IL: University of Illinois Press.

Spence, L. K. (2015). *Knocking the hustle: Against the neoliberal turn in Black politics*. Brooklyn, NY: Punctum.

Sterne, J. (Ed.). (2012). *The sound studies reader*. New York: Routledge.

Taylor, E. F., & Wheeler, J. A. (1992). *Spacetime physics*. Gordonsville, VA: W. H. Freeman.

Taylor, K-Y. (2016). *From #BlackLivesMatter to Black liberation*. Chicago, IL: Haymarket Books.

Tsing, A. L. (2005). *Friction: An ethnography of global connection*. Princeton, NJ: Princeton University Press.

Tsing, A. L. (2012). On nonscalability: The living world is not amenable to precision-nested scales. *Common Knowledge, 18*(3), 505–524. doi: 10.1215/0961754X-1630424

Tsing, A. L. (2015). *The mushroom at the end of the world: On the possibility of life in capitalist ruins*. Princeton: Princeton University Press.

Tuck, E., & McKenzie, M. (2015). *Place in research: Theory, methodology, and methods*. New York: Routledge.

Tyler, R. W. (1949). *Basic principles of curriculum and instruction*. Chicago, IL: The University of Chicago Press.

Whitlock, R. U. (2007). *This corner of Canaan*. New York: Peter Lang.

Winfield, A. G. (2007). *Eugenics and education in America: Institutionalized racism and the implications of history, ideology, and memory*. New York: Peter Lang.

Woodson, C. G. (1933). *Mis-education of the Negro*. n.p.

Young, M., Chester, J-L., Flett, B. M., Joe, L., Marshall, L., Moore, D., Paul, K., Paynter, F., Williams, J., & Huber, J. (2010). Becoming 'real' Aboriginal teachers: Attending to intergenerational narrative reverberations and responsibilities. *Teachers and Teaching, 16*(3), 285–305.

4 Policing Deafness

Everyday Sonic Oppression in Schools

We police children so hard in schools. This is not hyperbole. A central aspect of every teacher's role in P-12 schools is some version of a Prosecutor-JudgeJury combination in which teachers set rules, order spaceplaceperson through those rules, and mete out appropriate sanctions depending on the egregiousness of the infraction. But that, by no means, is all. Throughout the school day, we as educators take control of the timing and possibility for enacting their bodily functions, sustenance, movement, talk, information, and responses (e.g., Gershon, 2017; Gilbert, 2013; Mehan, 1979; Nespor, 1997; Woodson, 1933). We do so unabashedly and unequally (e.g., Cooper, 1892; Morris, 2015; Rist, 1973; Varenne and McDermott, 1998). As a result, children before the age for consent to drop out of school (often around 16 years of age) spend the biggest chunk of their day being manipulated, controlled, and suppressed. Note to teachers reading this chapter: No matter how amazing you are as a person, educator, or parent, you do this.

Does this help with safety for students? At some level, most certainly. Is there a benefit for getting the hang of how to share public spaces with others who are both like you and different from you? Absolutely, and one can come up with any number of possible reasons and rationales for policing children.

However, none of them is an excuse for the actual policing of children as if they are *de facto* criminals by actual police. Local reports of high school students being electroshocked with a Taser by police are frequent and, anecdotally, I can report that those incidents sadly tend to function as one might expect. Incidents are much higher at city schools than suburban or rural schools, and students of color are Tased much more often at suburban or rural schools than their Anglo peers. Similarly, boys are Tased more frequently than girls. If you'd like to check the veracity of this claim, please simply search for online images with some combination of the words: student, Taser, high school.

This is clearly policing run amok. Yet, it remains a possibility due to a particularly American combination of escalating weaponry, removal of children's rights in their roles as students, increasing police presence in schools under the guise of "resource officers," and the ease with which local police forces are receiving leftover or slightly used tools of war and their not so subtle trickle-down effect into schools. As but one of many possible other examples, although recent attention to "restorative justice" practices in schools are indeed a much more humane, respectful, and compassionate pathway for conflict and justice, such processes as practiced in schools are most often derivations of those used in the criminal justice system—a move that makes parallel children's behavior and those of criminals.

The reason I've begun the conversation here is that I find that, far too often, critical scholars in education make claims about control, policing, discipline, and the like without attending to either actual policing in schools or the kind of violence that is done to students through school-based police. There is much work to be done on this front, and I'll return to aspects of sound, education, race, and violence of this sort, particularly in Chapter 8.

There are at least two central concerns with non-police policing that happens in schools, actions that are often enacted by teachers under the guise of management or discipline that speak to the variety of possible reasons and rationales suggested in the previous paragraph. First is an argument that is often (falsely) framed as conditions for "real schools." To be blunt, actual schools are often overcrowded, under-resourced, underfunded, and poorly designed. Although this is certainly not always the case, it is common enough that a new, beautiful, school is a cause for celebration, particularly in school districts with majority communities of color—rare for elementary schools and even more rare for high schools. Additionally, high schools are often designed by the same companies that design prisons, with many of the same physical features: an open entry way with a high bridge where administrators can monitor students as they enter and leave for example. From this perspective, options for less control are not sensible because of the generally poor conditions in which we insist our children spend the majority of their childhood.

Second is an understanding of the kinds of social and cultural conditions under which someone can enact their sociocultural roles as "student" well. Contemporary schooling is in many ways governed by scholarship that touts the importance of time on task (e.g., Farbman, 2015), the need for exact modeling of procedure (White & Fredericksen, 1998) that are repetitive and generally predictable (Gershon, 2017), all of which make a lack of distraction or interruption of paramount importance in classrooms. Leaving aside the deep connections between such facets of schooling and business (e.g., Claessens, Earde, Rutte, & Roe, 2007) and how such environments conflict with the needs of children, what is clear is that in our test-driven curricular moment lessons need to move apace (Valli, Croninger, Chambliss, Graeber, & Buesse, 2008; also, for example, McNeil, 2000; Taubman, 2009). In order to cover the requisite material before annual tests that measure students' understandings, teachers' ability to teach, and schools' goodness, they need to do so regardless of students' needs or comprehension. For, without at least exposure to possible content, how will students possibly be successful on their assessments?

In this environment, as noted in Chapter 2, sounds are anathema to learning. But not all sounds. Teachers are to be heard loud and clear. Many newer schools have installed wireless microphone systems for teachers so that their voice can be heard throughout the classroom on multiple speakers inserted into the ceiling. To be sure, systems like these are helpful for teachers' health and students' ability to hear what their teachers say. They also accent the way sound works in classrooms—from teachers to students.

There is, however, an exception to this rule: deaf[1] schools. For those who have never spent time in a classroom where the majority population is deaf or hard of hearing, it can be a rather loud space. Yet, when speaking with deaf people about their experiences in majority hearing schools, what they share underscores the opposite perspective, a lack of resources and deep difficulties getting federally mandated services. The irony is profound. We want hearing kids to be deaf, deaf kids to be hearing, and both deaf and hearing kids to listen to their teacher.

Here, I attend to three central ways that sound manifests in contemporary schooling: sound policing, deafness as sonic deficit, and kicking ass through band. What these three aspects of the sonic have in common is their materiality, the ways in which sound physically affects ecologies and relations as well as their associated ideas. A brief section on the silences around Indigenous education and the histories of Indian boarding schools in the United States veers slightly into more metaphorical absences and silencing. However, lack of attention to this facet of U.S. curricular patterns and practices in some fashion on a chapter about othering, sound, and silencing would contribute to rather than interrupt these injustices. These arguments are then reassembled in the conclusion that wonders about the impossibility of hearing in schooling.

Sound Policing: Sound Suppression, Sound Aggression, and Echoes in the Hall

Children are meant to be seen and not heard, the maxim goes. What we want is classrooms that are places where students have unfettered sonic access to knowledge. We are concerned about outside sounds interrupting students' work, sounds of the city for example, and sonic disjunctures from inside, the sounds of children to each other interrupting their work and other sounds from within classrooms. What is almost never concerning are the sounds that teachers make. By this I do not mean simply the sounds of their voices. I've witnessed teacher's mandatory movement during annual state testing and the accompanying constant squeaking of their sneakers distract students; the loudness of the ticking of a clock causing children discomfort. These sounds, however, are teacher-made or institutionally driven and therefore do not count as the kinds of sounds that are unacceptable in schools.

Conversely, there are students for whom sounds are helpful. Those who have grown up in cities often find that the sounds of silence are in fact too noisy, the overwhelming quietude discomfiting the constant influx of mechanized daily experiences. Further, a general quieting of classrooms can create a context in which everyday sounds become disproportionately loud. Older air conditioning/heating units common in city schools across the nation can create a din beyond comprehension, for example, creating a context in which all conversation begins at a volume close to yelling and quickly crescendos to a point that secrets can be delivered through a conversational level of talk—cutting the talk does nothing to quiet the air conditioning unit.

Then there is the historical baggage of educational sounds that are reflected in contemporary fields of education and sound studies. In education, this understanding can be seen in everything from John Dewey's underlying assertion that natural learning in the out-of-doors is central to children living full lives, to assertions in normal schools that natural authentic education is equivalent to being in non-urban, outside environments, to various visions of place-based education that often have not-so-hidden biased valuation of "natural" ecologies. There is a parallel understanding in sound studies in those who work with acoustic ecology and continue to privilege "natural" environments while discounting "man-made" sounds (for more on this point, Feld, 1982, 2015).

The difficulty with this position from an educational perspective is twofold. First, it privileges an impossibility for most children in most schools at our contemporary moment. In so doing, it also places students and the sounds/movements in classrooms as antithetical to educational interactions. Second, it creates yet another iteration of dominant, Anglo understandings of volume. I say Anglo here to emphasize that the same groups categorized as racially non-white when they arrived to Ellis Island in the last century are often coded in the same sonic fashion. Irish, Italians, and Jews are all educationally loud. According to this typology, Latinx students are loud and become foreign when code-switching into Spanish. African American students, particularly girls, are the students who perhaps have their volume checked the most in school, their otherness in schooling pegged not only in the racial politics of optics, better melanin content, but also sonically (e.g., Fordham, 1993; Evans-Winters & Esposito, 2014). Where students of color are racially coded in ways that the sonic compounds the ocular, ethnically non-Anglo students who might otherwise pass can be sonically outed. When combined, these and other associated understandings create a context in which the sounds of students are suppressed.

Conversely, as I travel about the U.S. presenting sound-related papers, I am increasingly pulled aside with a rather similar tale. Educators who work at predominantly/ridiculously Anglo suburban/rural high schools, either as teachers or scholars studying in that context, share that a system, usually just a couple of speakers, has been installed in a particular hallway or two. These speakers broadcast a persistent high pitched noise that is out of the range of hearing for most adults and readily heard by high school aged students. Not so high pitched or loud as to be truly painful, the idea is that the sounds emitted are a constant, persistent discomfort for students, placed on particularly congested hallways and intended to cause them to linger less and generally disperse.

From one perspective, the halls are simply congested. Yet, in each of the over 10 cases I've heard about to date, the speakers have been placed in hallways where students of color most often congregate. In addition, there is some question about whether or not the hallways in question are actually clogged corridors. Instead, what I have heard about is that the hallways are, as one teacher told me, "I don't think they're that congested. They're also not too loud. What they are is the place where the Black kids hang between classes." As but one example, the following article deftly omits that the only corridor on which the "mosquito" noise is used is where the comparatively few students of color congregate: www.wthr.com/article/annoying-mosquito-noise-keeps-students-moving.

What is striking is that this has been the conversation each time I've been informed of this practice. While this isn't a big "n" quantitatively and is anecdotal evidence, it certainly points to a possible trend that is well-aligned with an understanding of sonic difference as social deficit. More disturbingly, it smacks of another conversation that I'm increasingly having at RAI (ridiculously Anglo Institutions) referenced at the beginning of this chapter, talk about the frequency and direction of students of color being Tasered in school.

When considered together, the use of sound as a deterrent for students of color is at once a reification of sonic differences as social deficits and a use of the tools for sonic warfare employed on students of color. Although the first instance may appear to pale in comparison to the second—the first is a sociocultural construction, the second a form of material harm—they form mutually reinforcing aspects of a feedback loop of aggression and violence. What makes this loop even more pernicious is that it is social differences that are most often utilized for the expression of academic deficits (for more on this, check Gershon, 2012). It is this feedback loop that constructs African American kids' everyday expressions as abnormal, directs the tools of sonic warfare at children, and then uses those two expressions as a means to further falsely explicate rationales that assert African American-ness as a foundation for academic anemia. It is also the case made against Latinx students with the added complications of language, a nuanced yet significant difference between being ostracized for speaking a recognized (yet still othered) English dialect and a language other than English, questions of immigration (real or perceived), and other such attending assumptions.

Importantly, these and other such factors lay out a clear foundation of an understanding central to schooling that is often either missed or addressed through a different trajectory. Being resoundingly othered is but another possible indicator that students of color—even without attending possible complexities of class, gender/identity/expression, sexual orientation, home language, mobility, and the like—are ontologically unacceptable in U.S. schools. The multiplicity of ways of being, the inherent complexities that are always greater within one group than between any two groups, are subsumed by the production of race into monolithic impossibilities of is-ness. By simply showing up to school ontologically themselves, an inescapable eventuality, students of color are necessarily always neither Anglo enough nor institutionally normal, a tendency that remains the norm even in schools where students of color are the overwhelming majority even as students in U.S. schools as a whole will never be this Anglo again (www.edweek.org/ew/articles/2014/08/20/01demographics.h34.html).

This, then, is the sound expression of systemic racism expressed in institutions through interpersonal interactions, in which all children should be quiet and Black and Brown children are silenced. It is an institutional racism that particularly impacts young people of color in their sociocultural roles as students while providing an officially sanctioned education (formal curriculum/official knowledge) that similarly silences African American and Latinx voices, histories, and knowledge throughout the curriculum (Apple, 2014; Gershon, 2017).

Should this sound like hyperbole, Mark M. Smith (2008) argues that slavery and the construction of race in the American South was grounded in sonic violence and oppression. Where slaves produced Black sounds often categorized as noise, poor whites, who had many of the same kinds of jobs and tendencies towards volume as enslaved African peoples, argued they were different by virtue of their white skin, a construction further enforced by more wealthy slave owners to sow sonic unrest that perpetuated the systemic and interpersonal horror that was slavery in the American South (M.M. Smith, 2008, pp. 5–10). Contemporary constructions of the sounds of Blackness as loud, noisy, and boisterous, is a continuation of antebellum era racist sensory politics. Should this somehow be in doubt, please remember that Jordan Davis, a 17-year-old African American high school student, was murdered by Michael Dunn, a 47-year-old Anglo software developer, because of what Dunn called, "loud music." Jordan Davis was killed for playing Black music at too Black a volume while being a Black teen.

Class in Session: A Brief Aside on the Sounds of Social Class

Although there has been an uptick of talk about social class, stemming in no small part from a recession in 2008 that has tended to right itself by further privileging the wealthiest and leveraging their wealth against an increasingly poor and numerous majority, discussions of how talk reveals and places students according to social class have been less numerous. There is little doubt, however, that how students speak, even when students are Anglo, that class distinctions in talk translate into classroom deficits (e.g., Horvat, Weininger, & Lareau, 2003; Lareau, 1987; Stuber, 2006).

Practically, this means that students of color can be further marginalized by elements of class in their talk and that Anglo students, in spite of social capital they might gain due to race, are again segregated by social class. When combined, this creates a context that in many ways mirrors Mark M. Smith's (2008) distinctions between poor white Southerners and slaves noted above. Where the ways that poor to working class Anglo students talk inscribes them with a sonic difference that is conceptualized as a sociocultural deficit, their Whiteness elevates poor Anglo's talk above students of color who might be able to transcend the social class of their speech but not the sociocultural construction of their race (e.g., Au, 2008; Delpit & Dowdy, 2008; Michaels, 1981).

Sensory Politics Goes to School: Deafness and Being Hard of Hearing as Sonic Othering

In their chapter focused on questions of deafness and the political nature of sociocultural debate, Joseph Michael Valente, Benjamin Bahan, and H-Dirksen L. Bauman (2011) wonder what happens "when a member of a minority culture violates the sensory norms of a dominant culture?" (p. 248). They assert that the violator is disciplined, controlled, and pressed to conform to dominant understandings. In this case, it is deaf people who are at once disciplined for making vocalizations outside of the norm, essentialized for their deafness rather than recognized for the multiplicity of people who are deaf, and underlines longstanding sociocultural politics of the senses, that deafness is a deficit rather than a fully functioning ontological possibility. Their argument documents how deaf people are doubly bound by the sonic, pressed by a phono-centric construction of the world in which deafness and deaf people are constructed as possessing/being a negative, absence of sound on one hand and through which deaf people's voices, when raised, are similarly inadequate, less than "regular" hearing voices.

> The argument of sensory politics is to foreground that there is nothing essentially better about corralling children to follow the socially dominant sensory order. This argument highlights the fact that decisions are often made with regard to the heavy weight of the social construction of the senses, of the role of voice as presence and the policing of sound and the interpretation of social class that ensues. . . . To the hearing world, the deaf person without voice appears to lack a sense of presence, identity, being. . . . In this light, we may see the alignment of education and speech as a particular cultural bias against sign language and alternative sensory worlds.
>
> (Valente, Bahan, & Bauman, 2011, p. 255)

That deafness is viewed as a deficit in educational institutions is not surprising. As John L. Luckner reminds readers at the onset of his UNESCO white paper on Issues in Education of Students Who are Deaf or Hard of Hearing, "Special education in the United States began with the opening of the American Asylum for the Education of the Deaf and Dumb (now the American School for the Deaf) on April 15, 1817 in Hartford, Connecticut" (n.p.). From a historical perspective, we in the United States spent the first 160 years of deaf education thinking of deaf children from a pathological perspective, "focusing on cures or mitigation of the perceived handicap" (Senghas & Monaghan, 2002, p. 69) and only the last 39 years considering deafness as an ontological category of being that includes conceptualizing the multiplicity of deafness among deaf and hard of hearing people.

There is no small irony in how deaf and hard of hearing students are treated in schools. In spite of schools having the legal responsibility to provide interpreters and other services for deaf and hard of hearing people (Meeting the Communication Needs of Students with Hearing, Vision, or Speech Disabilities, 2014), I have yet to meet the parent of a deaf or hard of hearing child or a deaf or hard of hearing person who has not had to fight tooth and nail for their legally mandated services. As but one example, in spite of having a rather strong sign/interpretation program at my current institution, deaf or hard of hearing students often rightfully complain that there are not enough interpreters to meet their needs.

In this and other such ways, there is a good deal of alignment between the double bind in which students of color and deaf and hard of hearing students find themselves in schools. Although both groups have a legal right to access an equitable education, SOC and deaf or hard of hearing students are supposed to "be" in schools in ways that are contrary to legally mandated possibilities for their participation. While critical scholars often correctly call our attention to the significant difference between fairness and justice—in which being fair is granting the same thing to everyone equally and justice is granting people what they need with attention paid to all levels of injustice and discrimination—here they are collapsed. In practice, deaf and hard of hearing students are granted neither the same rights as everyone nor the tools to compensate for inadequacies of the phono-centric world of schooling. Instead, they are twice passed over, given information in a singular fashion that is discriminatory (speech) and ostracized for their voice (speech), no small irony in a field that claims to value the voices of children and cling to long-since disproven learner-typologies, not as they were conveyed but instead as they continue to be poorly interpreted (e.g., she is a visual learner).

The degree to which deaf culture slips through the cracks in schools—largely absent from discussions of physical services that are needed to navigate a hearing school and missing from conversations about culturally relevant pedagogy—is further insult to ontology. Discussions of deaf culture are generally not part of teacher education programs, instead more often relegated to either a day or some readings in required special education courses or separate electives (Johnson & McIntosh, 2009). Then there is the push to cochlear implants, a move that at once presses against deaf culture as a significant and important aspect of being, that deafness can be a facet of society and culture without deficit, and a further iteration of hearing-as-normal (e.g., Sparrow, 2005; Valente, Bahan, & Bauman, 2011).

The point here is not that cochlear implants are not a viable option and I, as a hearing person with a hearing child, would absolutely not find it ethical to insist on a position to any deaf or hard of hearing person or parent of a deaf or hard of hearing child. Instead, the point is that, at best, a lack of attention to deaf culture and a strong tendency towards cochlear implants, and, at worst, an intentional othering of deaf or hard of hearing people, is a strong tendency in U.S. public schools. This is to say nothing of how deaf or hard of hearing students are painted with a monolithic brush that at once not only gathers them into a single category of deficit, deafness, but also does not address significant sociocultural differences within deaf or hard of hearing communities, between deaf or hard of hearing people of color and Anglo deaf or hard of hearing people, or differences in social class for example (e.g., Cohen, Fischgrund, & Redding, 1990; Valentine & Skelton, 2003).

Instead, what we most want is for kids to be hearing so that they, like all hearing students, can be quiet and hear the teacher. Although there are certainly many arguments on both sides of an ongoing debate about the place, use, and justice of cochlear implants and audism, teaching children to lip read rather than sign is an explicit manifestation of dominant, hearing sociocultural silencing of deaf and hard of hearing people, regardless of what other benefits it might indeed bring (for more on this from a sociohistorical perspective, check Bauman, 2008). From an educational perspective, deafness is still a disability, deafness remains largely unrecognized, and, just as is the case for students of color, the multiplicity of is-ness that is the infinite variety of how someone can be as a deaf or hard of hearing person, is subsumed by a focus on what is perceived as their ontological limitations. This remains the case even in instances where students' educational needs are met. Their sociocultural ways of being as a member of deaf or hard of hearing communities remains overlooked and minimalized through the very attention to meeting their physiological needs, bringing deaf or hard of hearing students back in a self-fulfilling prophecy in which their deafness cannot help but be constructed as deficit and their deafness is rendered mute. Where Black and Brown kids are silenced, deaf and hard of hearing kids are not given a voice, opposite sides of the same sonic coin.

Kicking Sonic Ass: Why Bands Matter in School

There is much talk about how music education programs help students build self-esteem, be more creative, and even improve important skills in other more "academic" areas like math with its similar attention to intervals, patterns, and numbers (e.g., An, Ma, & Capraro, 2011; Bahna-James, 1991). However, it has long been my contention that this is all secondary to why music programs in general and band programs in specific tend to outlast all other visual and performing arts in K-12 schools, especially junior high and high schools. Where upper grade elementary schools (grades four or five) used to be where younger students began their musical careers, after gutting school budgets and 15 years of overemphasis on annual standardized assessments, with a couple of exceptions,[2] the pathway now starts in middle school.

In short, there are three reasons that music programs tend to flourish when all other support to the arts often flounder. First, general band is the daily practice space for the kinds of music that students play in marching band, the halftime entertainment and expression of school spirit for high school football games on Friday nights throughout the United States. In addition, marching bands play the school fight song, interject musical "pep" and "spirit" over the course of the game, when their team scores a touchdown, and often serve as an additional set of de facto cheerleaders, hyping the crowd.

Second, in more affluent areas, communities with greater cultural capital, music and the arts are understood to be important values. While perhaps not the most charitable read of such contexts, it is important to remember that in the beginning of *Art as Experience*, John Dewey (1998/1934) asserts that wealthy groups and individuals in power value art to document that they care about something other than money, then use art to express their specific norms and values, and place the artworks in museums that are prohibitively expensive except for those with means. Affluent communities work that way in some respects, documenting that they care not just about academics but a well-rounded, creative person. However, wealthy communities also know that success in a visual or performing art can also look good on a college transcript to a more elite state or private school.

Finally, there is this: A band can kick another band's ass. Musical ensembles regularly go to local, state-wide, and national competitions where groups and individuals within groups compete against other bands and individuals. This is as much the case for instrumental groups as it is for vocal groups. Just like grades, groups are numerically ranked with overarching numbers (e.g., 1 out of 5) and composite scores from judges that function not too differently than deciding upon a valedictorian when more than one student has above a 4.0 average. As a result, musical ensembles can bring prestige to their schools in a way that art students in general cannot, by winning at sports-like competitions.

Yet, for all this competition and sound, music classrooms follow the same rules as non-musical classrooms. In spite of often having over 40 middle or high school students each with an instrument in the same space, music teachers often expect students to be quiet in between takes or when working with a section rather than that entire group. To be fair, this is also the expectation in professional ensemble settings as the quieter it is between takes, the more quickly the group can get to rehearsing. However, it is also somewhat unrealistic to expect 40 15-year-olds with instruments to sit silently. For example,

I'll never be able to forget Mr. Ball, my middle school band teacher, yelling, "Don't be sorry, be quiet!" at every band and jazz band class and rehearsal.

All this formal music education, however, runs contrary to informal music-making or sound generation. In these contexts, sounds, musical or otherwise, are relegated to not-learning or unwanted noise, the educational equivalent value of a jackhammer. Students are not to make noise in the halls during class times and, because there is always a class somewhere throughout the school day, music-making in the halls is prohibited, a prohibition that includes, if not especially geared towards, singing and listening to music on headphones.

There are, of course, some very good reasons why one might suggest that students not sing in the hallway, not least of which is that sounds of any sort can indeed interrupt classrooms, particularly during specifically quiet classroom moments, while students are taking a test for example. There are similarly good reasons not to listen to headphones during class. Aside from being generally rude not to more fully attend to a teacher or peer, students listening to headphones can miss important information from teachers and peers about the knowledge at hand. However, the overwhelming majority of scholars I know listen to music while they work and, of those who prefer to listen to music as they work, the majority often listen to music on headphones. There's something about the immediacy of sound, not to mention sound cancelling properties of having sounds piped directly into your ears by devices that sit in or on your ears, that is helpful if not comforting.

Why students aren't regularly provided the same opportunities is likely more due to the need for overarching rules for times when headphones should likely not be applied than it is for general use. There also remains a particular concern about students cheating on assessments, a now quaint-seeming idea that students will tell each other answers or have them programmed on their devices, a leftover worry from days of teaching past when students didn't walk around with computers in their pockets.

Both concerns are difficult to maintain when considered next to the following everyday iterations of student life in contemporary schools. It is true that some students are distracted with sounds just as with other aspects of life but an overwhelming amount of scholarship about noise and sounds in schools written over the past decade and a half was likely produced while the scholar in question was listening to music. It is also not unusual for students with various patterns that lie outside of mainstream notions of normalcy (rather than learning disabled) to have the ability to use headphones during class written into the legal documents that govern their care and learning while at school. Then there is an ongoing series of cultural shifts in mobile listening (e.g., Sterne, 2003, 2012) for which those of us in our 40s and early 50s were the first generation to come of age with now-antiquated-seeming Walkman and mobile CD players. It is important to note that, since our own Bye Bye Birdie moment (kids today. . .), K-12 and college students have never lived a life without mobile, personal sounds, have adopted wearing various sound devices as markers of cultural and social status, and have negotiated their own codes for communication within, around, and through mobile listening strategies.

Further, the use and presence of sounds is rather different in schools for the arts. For educators and people like me, walking into a school for the arts is a breath of fresh air. Hallways are filled with kids talking, singing, acting. Sounds travel from classrooms out into the halls and around the corner, this class acting out lines from a contemporary musical, that class working on their science projects. What is different is not the education students receive per se—students in public schools for the arts are to take the same annual assessments as their peers in non-arts schools—but instead the expectations that govern social expectations at schools for the arts. Arts school kids are supposed to be more extroverted, arty, and along with those understandings comes an increased tolerance for sound, as sounds are part of what make schools for the arts special. This does not mean, however, that schools for the arts do not have space for quiet, introverted students, for those students too are part of stereotypical arts students—the brooding, quiet, painter/author-type, for example. Yet, these sonic expectations are not solely the purview of arts schools, they are also applicable to two types of all-but-defunct schools, alternative and vocational schools, two kinds of schools that are, in their contemporary incarnations, places for "atypical" or "non-traditional" students, in spite of histories to the contrary (e.g., Kliebard, 1999).

This contrast in social policies of quietude between most public schools versus arts and vocational schools is another iteration of the first points raised above about the silencing of students of color and other non-Whitestream groups: Quiet is part and parcel of class and Whiteness. Quietude has been policed in noise-abatement programs, and those in the earlier years of the United States were some of the more successful programs in the world. These programs were successful in no small part because they linked the harm of noise not to social class and race but instead to potential harm to children at home and schools, leaving race and class distinctions more implied than stated (Bijsterveld, 2008): quiet as an upper class and Anglo, as well as supposed signs of intelligence, efficiency, and good work.

A Brief Aside on Policing Joy as Noise

Part and parcel of what is considered to be proper sounds in schools is tied to an Anglo sense of silence as golden and returns to a question of what is constructed as noise. In queering education, resonance, reverberations, and a collapsing of scale make possible room for all kinds of sounds that are often conceptualized as noises. Chortles and giggles and the snick of lips smirking; and the quick exhale of kinship across the classroom when only two of you get what's funny about that thing that happened; and the thought that boomerangs out of the recesses of your thoughts, careening across the surface and making you snort at just the moment the teacher asks if there are any questions; or when all the sound suddenly stops in the classroom just as you swear in a story to a friend that was completely clandestine until you say "shit" at a normal volume; or when you as a teacher sneeze in the middle of a point when the whole class is watching and incidentally fart at the same time in a way that is unmistakable to a class of seventh graders.

At their core, these and other associated sounds, a student pumping her hand and whisper-yelling "yes!" as the class reviews answers to an in-class assignment for example, are in many ways the sounds of joy. While having the whole class hear you swear as if you're yelling may not be joyful to you, it certainly brings a good number of students joy. A sneeze-fart is often mortifying for teachers but hilarious to students of all ages (I have allergies) and the resulting laughter, raspberries of lips trying to contain a giggle, and general effort it takes students to pretend like that event didn't just happen is joyful to students. These events are also moments in which the sonic interrupts the stoicism of everyday schooling norms and, in so doing, both provide potential relief from those norms and an instance of joy in school. Yet, far too often, rather than being a humanizing moment where a teacher makes a joke with a student for swearing or a teacher acknowledges that she (or he or they) too has bodily functions, they too are suppressed.

Their suppression is a triple maneuver. It is first predicated on an idea that sounds that are not the teacher working or other associated sounds of students working that can be heard in the silence of their actions, pencils scraping on paper, quiet conversations between group members, are noise. Second, one must include joyful sounds as antithetical to studious sounds, a move that instead of recognizing the significance of a joyful noise turns joy into noise. Finally, that noise is policed. Yet, in practice, it tends to operate in a much more direct fashion as the policing of joy.

Beyond Silencing: Indian Boarding Schools in the United States

Beyond the apex of educational silencing is the removal of Indigenous children from newly formed reservations, a silencing and violent move to be sure, to non-reservation boarding schools, designed to rid children of their "Indianness" as deeply and quickly as possible, regardless of the physical, mental, and social anguish it caused. The process moved from practice to policy at the turn of the last century, modeled after the Carlisle Indian School, run by Lieutenant Richard Pratt, he of the "kill the Indian, save the man" doctrine.

The Carlisle school was established in 1879 and used an inhumane social experiment in which Apache prisoners of war were shackled and taken to a camp in Florida where their hair was cut and they were forced to wear Western/European style dress and learn English and the sociocultural norms of the day. In spite of a comparatively large number of suicides, those who survived transitioned to non-Indigenous life (according to Pratt and others of his ilk). This then became the model for the Carlisle Indian School: Young children were taken from their homes on recently established reservations, taught Christianity, and had as many traces of their home culture, life, and history erased as firmly as possible. The centrality of Christianity had been a presence in Indian re-education since the Civilization Fund Act of 1819 that funded private Christian, missionary schools designed to "properly" education Indigenous people.

As K. Tsianina Lomawaima (1993) writes,

> American Indian children have been coerced or recruited into federally run off-reservation boarding schools from 1879 until the present. The publication of native memoirs and film and radio productions periodically incite public interest or excite public outrage, an outrage that occasionally translates into government investigations, reports (Kennedy, 1969; Meriam et al., 1928), and reforms. Scholars have detailed the history of federal educational policy and schools for American Indians, but only rarely have they portrayed boarding schools as arenas for a reciprocating exercise of power (Foucault, 1979, 1980; Sheridan, 1980) between school staff and students-in other words, as an interaction Indian students helped to create.
>
> (p. 227)

Here, Lomawaima articulates the ways in which young Indigenous women were further monitored and oppressed throughout their time at school and in spite of their many moves of resistance, as she explicates in a later article (Lomawaima, 1999).

> In truth, there was nothing natural or true about the tenets of colonial education: (1) that Native Americans were savages and had to be civilized; (2) that civilization required Christian conversion; (3) that civilization required subordination of Native communities, frequently achieved through resettlement efforts; and (4) that Native people had mental, moral, physical, or cultural deficiencies that made certain pedagogical methods necessary for their education.
>
> (p. 3)

The list of grievances is long, ongoing, embodied, and inexcusable (e.g., Adams, 1995; Fear-Segal & Rose, 2016; Lomawaima, 2000; Trazfzer, Keller, & Sisquoc, 2006).

Were there any doubts that education is indeed a continuing tool for settler-colonialism, in contrast to Canadian universities that are making at least *some* efforts to amend programs for both P-12 teacher educators and across fields and disciplines, discussions of Indigenous boarding school experiences are most often not part of P-12 teacher preparation programs in the United States nor does there appear to be a push for the inclusion of Indigenous perspectives at a national scale. Instead, U.S. education continues unabated, with occasional attention to African Americans, often as slaves and in a fight for civil rights, and even less focus on Latinx populations unless you live in the Southwest, a climate in which states like Texas continue to place knowingly racist and demeaning information about Latinx people into children's social studies textbooks (e.g., Wang, 2016).

Hearing Impossibilities: Silencing for Normalcy

Questions of silencing in schools are often conveyed in more metaphorical terms as a means to convey missing histories, ideas, and ideals (e.g., Weis & Fine, 2005). However, as documented in this chapter, such silencing is as literal as it is metaphorical. Associated with questions of classroom management and knowledge production in everyday classroom lessons, sounds are a tool for policing students and an indicator of knowledge control. Sounds mark Anglo students as normal, students of color as abnormal, and deaf and hard of hearing students as aberrant, constructions that are in keeping with histories of race and deafness in U.S. schools. That there is little scholarship about Indigenous talk in educational literature on schooling is as much about metaphorical silencing—the overwhelming absence of Indigenous people in school texts (e.g., Bryant, 2008) and missing conversations about the histories of Indian boarding schools in most schools of education—as it is about literal silencing, the absence of talk about significant matters such as poverty in tribal schools (Hermes, 2005). Given the centrality of stories and talk throughout Indigenous people in the United States "is an integral and highly prized part of the culture" (Webster & Yanez, 2007, p. 116), the absence of such talk in schools and schooling is significant. Conversely, the one group of students that schools tend to press into hearing are those who physically cannot, deaf and hard of hearing students.

In sum, we police sounds in two distinct yet intertwined aspects of the phrase. On one hand, the adults in schools, and children as their normalized proxies, work hard to control the kinds, amount, volume, tone, and tenor of school sounds. Policing of this fashion serves to reinscribe Anglos as those who "sound right" and understandings of students of color, Indigenous youth, and deaf and hard of hearing students as sonically deficient. The central difficulty in this construction is that, like the category of "at risk" (Gershon, 2012), it is impossible for students of color, Indigenous youth, and deaf students to sound right. If they speak with the sonic identifiers of their respective communities, they are deficient; if they work to compress their sound palates into the narrow confines of Anglo norms and values, their remaining sociocultural markers (race, class, Indigeneity, deafness or hard of hearing-ness) are still sonically remarkable. How many times have professors of color been told in their educational careers that they "are articulate" or "well-spoken" with the not-so-implicit tag of "for a (insert race here) person" so present that it doesn't need to be said aloud?

On the other hand, we police sounds in schools through questions of an imagined, acceptable volume that is not far removed from the kinds of noise abatement societies that were in fashion when the American curriculum was codified in the late 1930s and early 1940s during the Eight Year Study (Kridel & Bullough, 2007). It is an understanding of volume that is inescapably braided with notions of race and class in which loudness is connected to unacceptable behaviors and imbecility, and quietness, and the ability to be appropriately quiet, are signifiers of intelligence and moral fortitude.

Then there are deaf and hard of hearing students, those whose world is quieted to a point that one might think that they represent, if not the apex, then close to the top of the scale of acceptable sounded-ness at school. Except for the fact that deaf or hard of hearing students are a step too far for schools where educational boundaries are more frequently about audition than sight. This to say nothing of the sounds that deaf or hard of hearing students make as they sign or the ways that their speech is affected requiring more translational work for teachers and peers. Instead, we want deaf or hard of hearing students to be able to hear in order that they can be appropriately quiet.

It is therefore not quite enough to say that we want hearing students to be deaf but for teachers' voices and deaf kids to hear so that they can respond to their teachers' talk. This chapter strongly suggests that, for many students in American schools, being heard as normal is an impossibility. For a nation where the last Anglo-majority kindergarten class continues to recede into history and Anglo norms and values create sonic deficits for students of color, deaf and hard of hearing students, Indigenous students, and poor to working class students, this is no small matter indeed.

Our predilection towards silencing of students can be approached from the opposite direction: Listening isn't central to teacher's classroom roles. Instead of orienting towards wonder, an inquiry-centered process, we focus on answers. The point here is not that answers are somehow unimportant but rather that they need not, and in many cases should not, be the focus of educational interactions. It is also an orientation that begs an important question: Why is teaching about talking rather than deeply listening (Oliveros, 2005) to children?

Notes

1 Differences between deafness as a physical question of otology and Deaf culture is often delineated as I have done in this sentence. However, Joseph M. Vallente and Gail Bolt (2016) suggest that questions of identity including notions of deafness are "constructed in relation to the group and based on discursive categories that precede the individual," not to mention other complications about the within group diversity and how culture/physiology binaries can further reinforce an equally false binary of speaking/signing that again privilege spoken over signed communications (p. 346). To this end, they "use "deaf "throughout to acknowledge the diversity of deaf lives across contexts and lifespans" (p. 346) and I follow their lead throughout this chapter.

2 These exceptions tend to be either a) elementary schools that feed high schools with successful music programs or b) elementary schools in more affluent areas.

References

Adams, D.W. (1995). *Education for extension: American Indians and the boarding school experience*. Lawrence, KC: University of Kansas Press.

An, S. A., Ma, T., & Capraro, M. M. (2011). Preservice teachers' beliefs and attitude about teaching and learning mathematics through music: An intervention study. *School Science and Mathematics, 111*(4), 236–248.

Apple, M. W. (2014). *Official knowledge: Democratic education in a conservative age* (3rd ed.). New York:R outledge.

Au, W. (2008). *Unequal by design: High-stakes testing and the standardization of inequity*. New York: Routledge.

Bahna-James, T. (1991). The relationship between mathematics and music: Secondary school student perspectives. *Journal of Negro Education, 60*(3), 477–485.

Bijsterveld, K. (2008). *Mechanical sound: Technology, culture, and public problems of noise in the twentieth century*. Cambridge, MA: MIT Press.

Bryant, J. (2008). State Street: North Carolina and the Cherokee trail of tears. *Journal of American Education, 47*(2), 3–21.

Claessens, B. J. C., van Earde, W., Rutte, C. G., & Roe, R. A. (2007). A review of the time management literature. *Personnel Review, 36*(2), 255–276.

Cohen, O. P., Fischgrund, J., & Redding, R. (1990). Deaf children from ethnic and racial minority backgrounds: An overview. *American Annals of the Deaf, 135*, 67–73.

Cooper, A. C. (1892). *A voice for the South by a Black woman from the South*. Xenia, OH: Aldine Printing House.

Delpit, L., & Dowdy, J. (2008). *The skin that we speak: Thoughts on language and culture in the classroom*. New York: New Press.

Dewey, J. (1998/1934). *Art as experience*. New York: Perigree.

Evans-Winters, V. E., & Esposito, J. (2010). Other people's daughters: Critical Race Feminism and Black girls' education. *The Journal of Educational Foundations, 24*(1/2), 11–24

Farbman, D. A. (2015). *The expanding case for improving and expanding time in school: A review of key research and practice*. Retrieved from www.timeandlearning.org/sites/default/files/resources/caseformorelearningtime.pdf

Fear-Segal, J., & Rose, S. D. (Eds.). (2016). *Carlisle Indian Industrial School: Indigenous histories, memories, and reclamations*. Lincoln, NE: University of Nebraska Press.

Feld, S. (1982). *Sound and sentiment: Birds, weeping, poetics, and song in Kaluli expression*. Philadelphia: University of Pennsylvania Press.

Feld, S. (2015). Acoustemology. In D. Novak & M. Sakakeeny (Eds.), *Keywords in sound* (pp. 12–21). Durham, NC: Duke University Press.

Fordham, S. (1993). "Those loud Black girls". (Black) women, silence, and gender "passing" in the academy. *Anthropology & Education Quarterly, 24*(1), 3–32

Gershon, W. S. (2012). Troubling notions of risk: Dissensus, dissonance, and making sense of students and learning. *Critical Studies in Education, 53*(3), 361–373.

Gershon, W. S. (2017). *Curriculum and students in classrooms: Everyday urban education in an era of standardization*. Lanham, MD: Lexington Books.

Gilbert, A. (2013). *Courageous pedagogy: Enacting critical science education*. Charlotte, NC: Information Age Press.

Hermes, M. (2005). Complicating discontinuity: What about poverty? *Curriculum Inquiry, 35*(1), 9–26.

Horvat, E. M., Weininger, E. B., & Lareau, A. (2003). From social ties to social capital: Class differences in the relations between schools and parent networks. *American Educational Research Journal, 40*(2), 319–351.

Johnson, J. R., & McIntosh, A. S. (2009). Toward a cultural perspective and understanding of the disability and deaf experience in special and multicultural education. *Remedial and Special Education, 20*(10), 67–83.

Kliebard, H. M. (1999). *Schooled to work: Vocationalism and the American curriculum, 1876–1946.* New York: Teachers College Press.

Kridel, C., & Bullough, R. V. (2007). *Stories of the eight-year study: Reexamining secondary education in American.* Albany, NY: SUNY Press.

Lareau, A. (1987). Social class differences in family-school relationships: The importance of cultural capital. *Sociology of Education, 60*(2), 73–85.

Lareau, A. (2003). *Unequal childhoods: Class, race, and family life (with an update a decade later).* Berkeley, CA: University of California Press.

Lomawaima, K. T. (1993). Domesticity in the federal Indian schools: The power of authority over mind and body. *American Ethnologist, 20*(2), 227–240.

Lomawaima, K. T. (1999). The unnatural history of American Indian education. In K. G. Swisher and J. Tippeconnic (Eds.), *Next Steps: Research and practice to advance Indian education* (pp. 1–32). Charleston, WV: ERIC Clearinghouse on Rural Education and Small Schools.

Lomawaima, K. T. (2000). Tribal sovereigns: Reframing research in American Indian education. *Harvard Educational Review,* 70(1), 1–23.

Luckner, J. L. (n.d.). *Issues in education of students who are deaf or hard of hearing.* UNESCO White Paper. Retrieved from www.unco.edu/ncssd/resources/issues_dhh.pdf

McNeil, L. (2000). *Contradictions of school reform: The educational costs of standardized testing.* New York:R outledge.

Mehan, H. (1979). *Learning lessons: Social organization in the classroom.* Cambridge, MA: Harvard University Press.

Michaels, S. (1981). "Sharing time": Children's narrative styles and differential access to literacy. *Language in Society, 10*(3), 423–442.

Morris, A. (2015). *The scholar denied: W.E.B. DuBois and the birth of modern sociology.* Berkeley, CA: University of California Press.

Nespor, J. (1997). *Tangled up in school: Politics, space, bodies and signs in the educational process.* Mahwah, NJ: Lawrence Erlbaum.

Oliveros, P. (2005). *Deep listening: A composer's sound practice.* Lincoln, NE: Deep Listening Publications.

Rist, R. (1973). *The urban school: A factory for failure.* Cambridge, MA: MIT Press.

Senghas, R. J., & Monaghan, L. (2002). Signs of their times: Deaf communities and the culture of language. *Annual Review of Anthropology, 31,* 69–97.

Smith, M. M. (2008). *How race is made: Slavery, segregation, and the senses.* Chapel Hill, NC: University of North Carolina Press.

Sparrow, R. (2005). Defending deaf culture: The case of cochlear implants. *The Journal of Political Philosophy, 13*(2), 135–152.

Sterne, J. (2003). *The audible past: Cultural origins of sound reproduction.* Durham, NC: Duke University Press.

Sterne, J. (2012). *The sound studies reader.* New York: Routledge.

Stuber, J. M. (2006). Talk of class. *Journal of Contemporary Ethnography, 25*(3), 285–318.

Taubman, P. M. (2009). *Teaching by numbers: Deconstructing the discourse of standards and accountability in education.* New York: Routledge.

Trazfer, C. E., Keller, J. A., & Sisquoc, L. (Eds.). (2006). *Boarding school blues: Revisiting American Indian educational experiences.* Lincoln, NE: Bison Books.

Valente, J. M., Bahan, B., & Bauman, H-D. (2011). Sensory politics and the cochlear implant debates. In R. Paludneviciene & I. Leigh (Eds.), *Cochlear implants: Evolving perspectives* (pp. 245–258). Washington, DC: Gallaudet University Press.

Valente, J. M., & Bolt, G. (2016). The curious case of the deaf and contested landscapes of bilingual education. *Equity & Excellence in Education, 49*(3), 336–349.

Valentine, G., & Skelton, T. (2003). Living on the edge: The marginalization and "resistance" of D/deaf youth. *Environment and Planning, 35*, 301–321.

Valli, L., Croninger, R. G., Chambliss, M. H., Graber, A. O., & Buese, D. (2008). *Test driven: High-stakes accountability in elementary schools*. New York: Teachers College Press.

Varenne, H., & McDermott, R. (1998). *Successful failure: The school America builds*. Boulder, CO: Westview P ress.

Wang, Y. (2016). Proposed Texas textbook says some Mexican Americans "wanted to destroy" U.S. society. *The Washington Post.*

Webster, E. P., & Yanez, E. (2007). Qanemcikarluni tekinarqelartuq [One must arrive with a story to tell]: Traditional Alaska Native Yup'ik Eskimo stories in a culturally based math curriculum. *Journal of American Indian Education, 46*(3), 116–131.

Weis, L., & Fine, M. (Eds.). (2005). *Beyond silenced voices: Class, race and gender in United States schools* (Revised ed.). Albany, NY: State University of New York Press.

White, B. Y., & Fredericksen, J. R. (1998). Inquiry, modeling, and metacognition: Making science accessible to all students. *Curriculum and Instruction, 16*(1), 3–118.

Woodson, C. G. (1933). *The mis-education of the Negro*. Complete text online retrieved from www.historyisaweapon.com/defcon1/misedne.html

5 Students as Improvisers

The Extra-ordinary Negotiations of Daily Life in Schools

Students sit at their desks in their assigned grouped seating. All students have their Reading and Writing Workbooks open on their desks and are in some stage of filling in the blank lines in a sentence with an appropriate adjective. Mr. Gutierrez[1] *alternates between asking students what answers they provided to fill in a blank and turning to write the answer provided on the board along with the sentence in which that particular blank was located.*

"Yes, Donald? What did you have for number three?"

"Ummm, quiet?"

"Okay, quiet," Mr. Gutierrez replies turning to the board. "Lorenzooo . . .woke at . . . 5:00 A.M. and looooked . . . at the quiet street," he intones as he writes on the board. "Lorenzo woke at 5:00 A.M. and looked at the quiet street, at . . . the . . . quiet . . . street, good," he evaluates reading from the board then turns back to face the class.

Marcy sits with a piece of paper on top of her open journal writing what later proved to be a note to a friend as the lesson continues in this fashion. Kevin whispers quietly to James, both boys keeping their heads down while they fill in the appropriate blank. Isabella draws pictures of a person in a dress holding a flower on a piece of paper on top of her journal, foregoing writing in the blanks entirely.

Sitting in one of my usual spots at the rear of the classroom, I sense rather than see some kind of movement towards the middle of the room. A moment later, I locate its source. Josep has quietly slipped out of his seat and has managed to crawl about a third of the way across the classroom. Every time Mr. Gutierrez turns to write an answer on the board, Josep crawls a bit further across the classroom floor, occasionally tapping a friend's leg, tugging on a sleeve or tapping a backpack, Cheshire-cat like grin on his face as he travels. Other than a fleetingly stifled giggle or brief look of pleasant surprise, students give no indication of anything out of the ordinary. They too keep their glances towards Josep's progress and any indication of their awareness that anything might be out of the ordinary to those moments when Mr. Gutierrez turns to write on the whiteboard.

After reaching the last row of seats, Josep promptly pops up to stand at Mr. Gutierrez's desk at the back of the room opposite from where I sit. He rifles through the items scattered on the top of the desk making increasingly more

> *sound until Mr. Gutierrez interrupts himself, whirling around from the board to ask,*
>
> *"Josep! What are you doing at my desk?"*
>
> *"Getting a paperclip," Josep casually replies.*
>
> *"Get back to your seat!" Mr. Gutierrez demands.*
>
> *Josep complies more quickly than usual, returning to his seat amid the laughter of his peers, who then turn again to face Mr. Gutierrez as they begin to fill in the remaining blanks in the respective pages of their Workbooks.*

In most mainstream discussions of teaching and learning, Josep's choices here are likely to be categorized as off-task behavior, decisions that lie outside the purpose and direction of a given lesson that on their surface appear to be disregarded (Foley, 2010; Page, 1990; Shumate and Wills, 2010; Thornborrow, 2003). Similarly, while Josep appears to have worked hard specifically not to distract his classmates with his movements across the classroom floor, his movements at his teacher's desk would likely fall into the category of disturbing or interrupting classroom lessons. From a more critical perspective, Josep's decision to crawl across the length of his classroom floor could be understood as a way to cope with the routinized nature of daily classroom lessons, a reaction to boredom. Along similar lines, Josep's crawl could be understood as an act of resistance, of using his available power to negotiate literal wiggle room from a seemingly static, top-down, teacher-centered lesson (e.g., Diamondstone, 1999; Langhout, 2005; Mirón & Lauria, 1998; Tuck & Yang, 2014).

Josep does indeed appear to be following none of the explicit or implicit rules of this fourth grade classroom. For example, he is not working in his workbook, attentively listening to his teacher or his peers, and is most certainly not in his seat. Yet, Josep is able to get across the entire length of the classroom without disturbing the flow of the lesson Mr. Gutierrez is conducting, a maneuver that seems to suggest no small understanding of classroom participation structures, delivered curriculum and pedagogy, and academic content.

This raises several important questions. What does Josep's skillfully executed crawl reveal about the classroom, daily lessons, and his relation to both? What means for conceptualizing classrooms could articulate both the processes that govern those classroom interactions and *students'* daily negotiation of these often-implicit interactional expectations? What way of considering students' classrooms interactions could capture the intended as well as unintended lessons students learn? How might students' classroom interactional choices be conceptualized in a way that they can be understood as various kinds of agency inclusive of yet not limited or reduced to off-task, subversive, or resistant behaviors? In what way might students' daily classroom interactions be conceptualized that posits them not only as reactionary but also as active negotiators of spoken and embodied classroom participation structures?

These questions are important because they speak to the very nature of what it means to be a student in classrooms and the kinds of knowledge students learn during daily lessons. This kind of complexity requires an equally flexible means for conceptualizing the complications and nuances of being a student, one that can support the cognitive dissonance of the possibility that Josep is simultaneously off-task, resistant, and bored or distracted, a distraction, and yet integrally involved in the lesson, critical of his context, expressing himself creatively, and a kind of relief, albeit comic, for his peers.

The answer that I propose here is a reconceptualization of students as improvisers. On one hand, such a conceptualization requires little discussion. As scholars such as those in Hallam & Ingold's (2008) edited collection have documented, the daily life of social actors is a constant improvisation, movement within, through, and around sociocultural frameworks for interaction, thought, and behavior.

On the other hand, improvisation continues to be a rich means for discussing various aspects of education. Such examinations tend to focus on teachers' interactions, improvising as a learning tool for particular groups, and/or classroom moments of improvisation (e.g., Borko & Livingston, 1989; Erickson, 1982, 2004; Gutierrez, Rymes, & Larsen, 1995; Lobman, 2006; Mitchell, 2008; Sawyer, 2001). As a result, students tend to be cast as respondents to teacher's improvising. To date, discussions of improvising in classrooms present improvising as moments within more structured classroom interactions such as the classroom participatory structures described below (e.g., Sawyer, 2011).

There is good reason for this perspective. Unlike teachers, students are continually on the receiving end of someone else's plan. In their sociocultural roles as students, it is not only their jobs to be open and ready to take on what their teacher has planned for them, it is expected. While there are certainly patterns for classroom participation structures as well as general norms and values, as anyone who has ever been a student can confirm, each teacher has her own nuances, some less obvious than others. However, unlike previous discussions of students' improvising (e.g., Gershon, 2006; Erickson, 1982; Sawyer, 2011), I understand improvising not to be the passing moments of interaction but instead to comprise the corpus of students' interactions in their sociocultural roles as "students."

Conceptualizing all student interactions as improvising provides a complex, critical means for considering how students enact their sociocultural role in order to better understand what happens in educational contexts like classrooms. Due to its location at the intersection of aesthetics, politics, theory, and practice (Bailey, 1993; Lewis, 2008; Montuori, 2003; Sawyer, 2001), understanding students as improvisers both captures the dissonant nuance of students' daily interactions and can be used to examine interactional and other processes that students engage as they improvise.

Perceiving students as improvisers is important because students' face-to-face interactions are moments of feeling, thinking, interacting with complex ecologies, and teaching. As has been documented over and over again in sociocultural examinations of classrooms, static frameworks for understanding teaching and learning tend to simultaneously reify dominant ways of knowing and being at the same time they reify traditionally marginalized students and teachers' status on the margins (e.g., Cooper, 1892; Valencia, 2010; Valenzuela, 2005; Valenzuela, Prieto, & Hamilton, 2007; Woodson, 1933). Additionally, and seemingly in contradiction to contemporary schooling's focus on standardizing academic content and its delivery, there is strong evidence to suggest that static lesson delivery, and similarly static means for understanding classroom interactions, negatively impacts all students, and some more than others (e.g., Gershon, 2017; Alonso, Anderson, Su, & Theoharis, 2009; Anderson, 2009; Mehan, 1979; McNeil, 2000; Valenzuela, 2005; Valli & Chambliss, 2007).

Of similar significance, educators of all stripes tend to focus on what students do in classrooms as "learning." While this is certainly a central part of what people do in their sociocultural roles as students, it is by no means the sum total of what students do in educational contexts such as classrooms. Although I have made this point in passing and focused on it more completely elsewhere, it is central to the significance of conceptualizing students as improvisers that I address in this chapter.

With few exceptions, a teacher's interactions are considered to be teaching—the verb for the role. What students do, on the other hand, is often reduced to questions of learning. From this perspective, students are either on- or off-task, engaged in interactions that can be observed and catalogued as learning or not learning. The difficulty with this perspective is twofold. First, as curriculum scholars have long demonstrated (e.g., McCluskey & Smith [Bethune], 2001; Cooper, 1892; Apple, 2014; Jackson, 1992; Flinders, Noddings, & Thornton, 1986), students learn not only the formal, intended curriculum but also implicit, hidden curriculum; not only the official curriculum but the unofficial curriculum; not only what is said but the null curriculum that is intentionally absent (e.g., Flinders, Noddings, & Thornton, 1986).

Second, and of equal importance, what students do in classrooms may be important to them but may not be what they are intended to learn or, perhaps, may not be "learning" at all. For example, as I detail below, there are several students in the opening vignette who are following the motions of the lesson's academic and social participation structures, writing down the correct answers at the correct time, yet who are doing so specifically in such a way that there can be a strong argument made that they *not* learning the intended lesson (formal curriculum). Yet, their teacher does not negatively sanction this group of students because they are going through the surface motions of learning. Similarly, it is very difficult to mark the exact moment when someone does learn or to somehow control for all the factors that might otherwise contribute to a moment of learning—a previous conversation or a sudden inspiration at a sound or smell for example.

For these reasons, I call what students do in classrooms "studenting" (check, Gershon, 2017). This term is useful both because it a) opens up the possibilities that what students are doing may in fact be one way for them to enact their role as students even if it is not learning and b) because it is a useful reminder that what students learn and what they are intended to learn, even when they might be performing a given task in exactly the way their teacher envisions, are not necessarily one in the same.

Conceptualizing students as improvisers is therefore significant in that it provides a more fluid and facile means for understanding the intricate interactions that constitute studenting. In its ability to both reveal the implicit structures that render students' improvising sensible and examine students' interactions in relation to those frameworks, understanding students as improvisers can also be a way to explore questions of educational equity and access, how contexts constrain local actors and their available agency for negotiating those contexts towards more socially just classrooms and processes. Additionally, conceptualizing students as improvisers provides the space to consider what students do in classrooms according to more student-centered constructions of sense-making—what makes sense to students is not necessarily as sensible to teachers and vice versa. Utilizing the opening vignette as a springboard, this chapter enunciates the ways in which understanding students as improvisers provides a means for making sense (Gershon, 2011; Ellsworth, 2005) of students' intended and unintended meanings as well as a means for interpreting the social and academic structures of classrooms and students' negotiation of these structures in daily lessons—from the curricular meanings they construct to their available interactional choices.

To these ends, this chapter has been divided into the following sections. I begin with a review of the literature on improvising as a metaphor in, and to some degree out of, education. These understandings tend to focus on adults as improvisers, moments of improvisations, or improvising as an educational tool for specific student populations. In order to provide a sense of the many ways in which studenting might be conceptualized as improvisation, the second section draws upon various traditions of musical improvisation to illustrate the critically creative skill and knowledge necessary for improvisation. This is important because it not only expands upon the more commonly utilized jazz improvisation metaphor but also because by laying these traditions side-by-side one can begin to see the depth and breadth of possibility in understanding studenting as improvisation.

Finally, in light of the many sociocultural constructions of improvising across and within musical traditions across the globe,[2] it is also important to note that the four improvising traditions presented here, and my focus on three of these iterations, are meant to serve as examples for the depth and breadth of improvising traditions towards an argument that all student interactions are improvisations. As such, this chapter is not intended as an exhaustive exploration of improvising, improvisers, or improvisational tradition. This said, if criticality is understood as an orientation towards and explicit attention to questions of power, both the chapter and this book are critically oriented and lean towards what has come to be known as "critical improvisation studies" (Lewis & Piekut, 2016a; Lewis & Piekut, 2016b).

The last two sections return to the opening vignette in order to provide a concrete example of how studenting is improvising. In the first of these two sections, I provide information about the contexts that inform students' interactions in the opening vignette. Drawing upon the improvisational models provided in the second section, the latter of these two sections presents students as improvisers through three different lenses of musical improvisation. A final brief concluding section revisits the notion of students as improvisers and the importance of such lenses for conceptualizing what students do in classrooms.

Classroom Interactions as Improvisation

Discussions of educational improvising generally utilize improvisation as a lens to describe at least one of the following three categories: particular kinds of classroom interactions or moments (e.g., Lobman, 2006; Varenne & McDermott, 1998), the teacher-as-improviser (e.g., Sawyer, 2004; Sharma, 2008), and improvising as a sociocultural bridge to learning for traditionally marginalized school populations, particularly African American teachers and students (e.g., Foster, 2001; Hanley, 2007; Mitchell, 2010).

The overwhelming majority of scholarly uses of improvising to discuss educational contexts share a construction of improvising out of the American jazz tradition (e.g., Dance, Gutiérrez & Hermes, 2010). Considerations of improvisation in such areas as organizational theory (Meyer, Frost, & Weick, 1998; Montuori, 2003) also frame their constructions of improvising on a particular kind of jazz soloist model—a soloist does the bulk of the improvising with fluid support, in both maintaining and improvising a song's harmonic and rhythmic structures, from the non-soloing musicians. This kind of support also jibes with Sawyer (2004) and Lobman's constructions (2006) of the ways in which improvisers in theater (in Sawyer's case, particularly out of the Second City school of comedic/sketch improvising) work with one another to make meaning—through agreement and support of the speaker's direction/ideas.

As has been noted elsewhere, while this kind of jazz-related improvising can indeed inspire critically creative movement within and away from implicit and explicit structures (interactional, intellectual, organizational, pedagogical, etc.), they can also serve to limit both what might be considered improvisational and how such improvising might be conceptualized (Kamoche, Cunha, & Cunha, 2003; Lewis, 2008; Zack, 2000). In light of the ways in which "improv" (improvisational theater "games") tend to rely upon often linearly sequential, intentionally consensual choices as both Sawyer and Lobman describe, these critiques are also in many ways also applicable to iterations of theatrical improvisation.

There is comparatively much less scholarship on students as improvisers. While there are both claims about the possibility (and importance) of understanding students as improvisers and empirical studies of students improvising (Erickson, 1982, 2004; Pinar & Irwin, 2005; Sawyer, 2001), such scholarship tends to either note how students are improvising a) within a broader framework or b) in response to their teachers' interactions. Because of such framing, discussions of students' improvising tends to be bounded either within broader discussions of moments of improvisation or closely tied to discussions of both the musical nature of students' talk or their reactions to adults' conversations. Ultimately, then, students' improvising, similar to teachers' improvisations, tends to be conceptualized as frames, bounded interactions that although part of everyday classrooms are somehow still set apart.

There is, however, a good deal of scholarship to suggest that daily interactions are indeed governed by socioculturally constructed norms and values that combine to form often implicit (yet no less real in their implications for people's lived experiences) expectations about how one is to interact with others, an understanding about the nature of human interactions both within and outside of educational contexts (e.g., Cazden, 1988; Kondo, 1990; Ortner, 2006; Spindler & Hammond, 2006). Erickson (1982, 2004) and others (e.g., Cazden, 1988; Mehan, 1979; Page, 1991) refer to the ways in which classroom actors work together to simultaneously recreate and innovate within their daily interactions as classroom "participation structures."

It is important to note that improvisation is often related to questions of race and culture, and rightly so. In an American context, this is largely due to the improvisational nature of jazz, a tradition that while expressed by musicians of a multitude of races, ethnicities, cultures, and nations, and is similarly performed by women, men, and members of the LGBTQ community, nonetheless has distinctly African roots. The same can be said for hip hop. As a result, a good number of the articles on teaching and improvisation address improvising as it relates to African American teachers and students, again, and rightly so.

This said, as briefly enunciated in the introduction, my argument here is that all student interactions—regardless of educational context (in or out of formal schooling), student or teacher sociocultural markers such as race, class, gender, sexual orientation, etc., interactional structure (formal/informal, implicit/explicit), or local/less local norms and values—are improvisations. I understand this argument to be inclusive of scholarship that has illustrated the potential for improvisation for African American teachers and students. This is a both/and rather than an either/or construction of improvising.

It is my contention that rather than creating a troubling universality of perception—as there are important racial, cultural and other such differences in approaches to improvising (for example, Lewis, 1996, 2008)—conceptualizing students as improvisers creates the space for multiple (even simultaneous) understandings of how people enact their role as students.

Epistemologically and empirically speaking, conceptualizing all students as improvisers and all student interactions as improvising creates the space for a given student interaction to be seen as simultaneously proactive, off-task, educational, resistant, and revealing of local structures, interactional and otherwise. It similarly provides the opportunity to focus on the ways in which sociocultural precepts intersect with and otherwise inform students' improvising.

In sum, understanding students as improvisers more clearly conveys the complexity and nuance of students' minute-by-minute interactional decisions (see Erickson, 2004) that are often overlooked for discussions of how those decisions fit within adult-centric constructions of educational purposes and meanings (for example, Thorne, 1993). Improvising is a skill that requires knowledge of both content and context, an awareness of both frameworks of understanding and the explicit and implicit possibilities those frameworks contain, constrain, and enable. Furthermore, conceptualizing students as improvisers asks the adult educators who examine studenting to not only address their own adult-centric perspectives but, importantly, also to wonder what such interactions might mean to students outside of such adult-centric perceptions of students' classroom choices. Again, understanding what studenting means to students in a given interaction provides an understanding of what students are learning (as opposed to what they are imagined they should be learning) and is inclusive of interactions that may not be learning per se (daydreaming for example).

In order to demonstrate that all students are improvisers requires both an expansion of mainstream notions of improvising and the applicability of this expanded frame to students' interactions. To these ends, the following section first expands upon the critiques of what I call "straight ahead" jazz improvisation including those briefly presented in this section. I then apply this expanded understanding of improvisation to the vignette I have come to think of as Josep's epic crawl introduced at the onset of this chapter.

On Improvising In and Out of Music

Improvising relates not only to "spontaneous composition" or Ornette Coleman's (1983) famously freeing or obtuse (depending on one's sensibilities) discussions of harmolodics but also the action of creating the space for improvising as well as "doing anything on the spur of the moment" (http://dictionary.oed.com/cgi/entry/50113553?query_type=word&queryword=improvise&first=1&max_to_show=10&sort_type=alpha&result_place=2&search_id=K7jC-1ORd2U-17122&hilite=50113553). Although these understandings still often pertain to the sonic nature of their origins (relating to music and talk), they have also, as these definitions describe, come to be applied to more spontaneous sociocultural choices outside of sound. More recently, as Lewis & Piekut argue (2016), the moment-by-moment basis in which social actors negotiate often-implicit interactional ideas and ideals applies as much to the ability to create an occasion as to create *for* an occasion, critical questions across fields and disciplines.

In light of such understandings, this section examines what improvising means both within and outside of musical interactions and is divided accordingly. The first subsection addresses multiple understandings of what improvising might mean and how it can function, setting dominant Western notions of improvisation against some forms of non-Western improvising as well as more marginalized Western forms of improvising. The latter half of this section considers how improvising has been theorized in discussion of human interaction and meaning-making outside of (though not necessarily separate from) musical traditions.

It is important to note here that there are other artistic forms that rely upon improvisation, particularly in theater. While these discussions of improvising tend to fall outside the scope of my argument here, it should be understood that such constructions of what it means to improvise can and should be utilized to further expand the conclusions I reach below. Of perhaps even greater importance is the acknowledgement that—because what counts as "structure" and "improvising" are social constructions based on interwoven layers of local and less local norms and values—most talk about improvising, musical and otherwise, are those understandings that most resonate with Western constructions of these meanings. Although I address scholarship in which notions of improvising are interrupted in the second subsection below, there is also much that could be added in this regard, additions that would serve to simultaneously enrich and critique current understandings of studenting, the broad range of interactions in which students enact their sociocultural roles as students, and in conceptualizations of improvisation both in and out of musical contexts.

On Musical Improvisation

Musical improvisation can be understood as movement that works either within or outside of a given musical framework. Even though jazz improvisation, the music most associated with improvising in (U.S.) educational literatures, is spontaneous in expression, musicians dedicate a great deal of time and effort practicing how to negotiate particular musical patterns in an equally particular manner (Berliner, 1994; Fischlin & Heble, 2004; Groesbeck, 1999: Lewis, 2008).

Although there are many permutations to this way of performing jazz, and the following nomenclature is not the cleanest fit for these multiple interpretations, I refer to this kind of jazz here as either "jazz" or "straight ahead jazz," terms that are often used interchangeably in the literature (e.g., Berliner, 1994). Additionally, as I have noted elsewhere, the majority of work in and out of education that makes comparisons between jazz and their field tends to rely on this understanding of what jazz means and how soloing functions within this musical framework.

In straight ahead playing, a group plays a song so that the melody is largely recognizable either once or twice through. Members of the group then take turns improvising on that melody. To do so, they utilize the chord structure of the song, the progression of multiple notes that help make the melody sound melodic, as the framework for their improvising. As Steve Coleman has noted (http://members.m-base.net), in spite of a deeply intertwined expression and consideration of rhythm, there is often an overwhelming emphasis on melodic over rhythmic improvisation in contemporary jazz, particularly for those outside of the "rhythm section" (piano, [guitar], bass, drums).

In this form of jazz, one musician usually takes the improvisational lead, while the remaining group members either stay out of the way by not playing or alter their musical choices to support, spur on, or challenge the improviser. Although these are indeed forms improvising, they are generally not the lead to a point that those not soloing often work hard not to "step on" the lead improviser's lines. At the same time, straight ahead jazz is never so straight ahead. It is nuanced, complex, and improvisatory. For example, musicians play inventive lines rather than the actual melody, catch bits and pieces of sound pitched by other performers, and turn them into platforms for further improvisatory play, and make tens if not hundreds of moment-by-moment decisions that alter, enhance, and otherwise engage in musical interactions that can be understood as improvisational.

In order to do this kind of improvising, jazz musicians must know the melody and the chord structure. This requires a deep knowledge of music theory, scales, modes, notes, and tones, as well as technical skill on one's instrument honed by years of practice (cf. Berliner; Monson, 1997). Thus, in a straight ahead model of improvising, soloists negotiate available musical frameworks on a moment-to-moment basis. As other group members adjust in turn to the soloists' decisions, the choices soloists make in turn affect the musical structures that inform their improvising.

This is in several key ways an oversimplification for the intuitively present, emergent sonic meanings and non-meanings that combine to create a straight ahead jazz performance, something I have experienced countless times both on and off stage. As I move into the following section, the point I will be underlining here focuses on the limitations of this construction as a model and metaphor for non-musical interactions. In short, while my aim is to provide a particular critique of a formulaic way for moving through musical possibilities that are literally endless—and as importantly, the troublesome ways that this kind of improvising is framed to describe educational contexts—this is by no means intended as an argument against straight ahead jazz or the talented musicians who play it.

It is, however, a critique of its use to describe education and is aligned with other such critiques, musical and non-musical alike, that I present in the following subsection. This critique is of playing straight ahead with a rigid adherence to a song's compositional, tonal, and rhythmic possibilities.

From a particular strand of this kind of playing, and the kind often taught in K-12 schools and universities, there are: "right" and "wrong" notes, notes that are too much against the grain of a particular chord structure, tones played only in passing from one set of sounds to another, and a rhythmic sensibility that needs to be maintained throughout an improvisation.

Again, the point here is not that music does not have rules—many that are much narrower and well-defined than in playing straight ahead—or that such rules do not help musicians communicate or work towards a common goal of sonority. Rather, it is that the kinds of playing that are associated with high-brow jazz, a milieu that has become parallel to Western art music as jazz has become "America's classical music" (for example, the arguments levied at Ken Burns' *Jazz* documentaries), has in many ways become canonized and routinized. It is these kinds of ideas that are often overlooked when educational scholars talk about the openness and freedoms of improvising.

Western Improvising Outside of a Straight Ahead Jazz Model

Although straight ahead jazz is certainly creative in its expression and critical in its process, the need to largely work within a framework can also be limiting, particularly when applied to non-musical contexts (Kamoche, Cunha, & Cunha, 2003; Zack, 2000). For example, in an article about the possibility of understanding classrooms through a lens of collective improvisation, I noted that such limitations are also present when straight ahead jazz metaphors are adopted into other fields such as education. There, I argue that one difficulty of applying a straight ahead jazz metaphor to classrooms is that the teacher is often cast in the soloist's improvising role while students are framed as accompanying musicians whose central responsibilities lie in responding to the teacher/soloist's choices.

In its stead, I suggest that viewing classroom interactions as collective improvisations might better enable observers to see where students regularly exercise their available power through meanings and possibilities that are emergent as often as they are structured (e.g., Davis & Sumara, 2006; Doll, Fleener, Truiet, & St. Julien, 2005). In their critique of a singular vision of improvising as it is applied in the field of organizational theory, Kamoche, Cunha, & Cunha (2003) make similar claims.

> [There is a] danger of jazz becoming what Weick (1980) calls a 'blinding spot', by obfuscating contributions from other areas of human endeavour and, most importantly, from grounded and empirical research. It is important not to be mesmerized by the jazz metaphor to the point of ignoring the potential contribution of alternative metaphors and other avenues of theory-development in organizational improvisation.
>
> (p. 2024)

Zack's (2000) response to the special issue of *Organization Science* on jazz improvising and organization (Meyer, Frost, & Weick, 1998) resonates with both of these perspectives. On one hand, he argues that, "overemphasizing the structure of traditional improvisational genres may limit the power of using jazz as a metaphor for innovation" (p. 228). On the other, under the heading "playing 'outside' may be the truest form of improvisation" (p. 229), Zack contends that,

> Variation is the stuff of bebop—extending the notion of harmonic structure and the rules for picking good notes. Improvisation, then, would refer to the maximal innovation that comes from improvising the entire composition spontaneously: its premise, its harmonic structure, its tonal language, and the actual sounds played.
>
> (p. 230)

Lewis' (2008) widely acclaimed work on the Association for the Advancement of Creative Musicians (AACM) notes how associated musicians have pushed at the boundaries between structure, improvisation, composition, and interpretation. In this incredibly thorough and engaging history of one of the most overlooked and influential (and ongoing) musical movements in America, Lewis reminds the reader of the depths and possibilities in, of, and for improvising.[3]

In short, improvising can serve as the source of material for musical composition; be utilized by players to move within and outside of compositional structures; constitute musical spaces, places, and identities; serve as the means for increasing musicianship in practice and pedagogy; be used as a pedagogical approach to teaching music; and as a metaphorical approach to all interactions.

Lewis also clearly demonstrates the difficult sociocultural/political/economic space in which the African American musicians that became the AACM found (and to no small degree continue to find) themselves. Because of members' interests in creating new music that was neither "jazz" nor adhered to formal conventions of Western art music ("classical") they were largely ostracized by both communities, a double-bind that resulted in the founding of the AACM, a place where Black musicians could meet, learn from one another, promote the music they created, and explore ideas and ideals. Now just past its 50th year, the AACM (www.aacmchicago.org) continues to move forward in its openness towards both composition and improvisation as 50th anniversary celebrations indicated.

As Lewis details, the American AACM rejected the terminology "free jazz" to describe their music. They did so because the term was associated with European improvisers whose approach to music was to be completely free without structure, where the AACM was and remains open to music regardless of its structure or history (including Western art music), and because free jazz musicians were often subsidized and supported by governmental arts associations (Lewis, 2008, pp. 215–258).

In light of these differences, I use the term "collective improvisation" to invoke the kinds of all-is-possible kinds of improvising associated with the

AACM. The use of this terminology also speaks to the usually interactional nature of classrooms and makes a connection to previous scholarship about the kinds of improvising that happen in classrooms (e.g., Erickson, 1982). It is important to note that the term collective improvisation is akin to the term "open improvisation" or simply "improvisation." Where all three terms (collective improvisation, open improvisation, improvisation) refer to music that is previously unscripted, as their names indicate, "collective improvising" refers to open improvising that involves more than one person where the terms "open improvisation" or "improvisation" encompasses the possibility of solo expression.

As the music of the AACM indicates, there are Western notions of what improvising means and how it can function that encompass yet supersede straight ahead jazz's notions of improvisation, as movement both from and in relation to prescribed structures. There is a world of freely improvised music that often gets (perhaps poorly) categorized as more experimental forms of jazz, due to its improvisational nature, or Western art music, due to its connections of an "avant garde." In practice, while they certainly have aspects of both, these forms of music are not necessarily either jazz or "classical" but instead forms of musical improvising that may or may not utilize jazz vocabularies.

For example, Pauline Oliveros' long history of improvising and her development and use of what she calls "deep listening," speaks not only to multiple musical traditions but also to the depth, complexity, facility, and beauty in and for musical improvisations. There is much to be gleaned from Professor Oliveros' lifetime of work in general and her construct of deep listening in particular, understandings that I take up in the concluding chapter to this book. Oliveros', Lewis', and members of the AACM's work also demonstrates how those who are adept master improvisers are also strong composers in both traditional and less traditional compositional forms (see also, for example, the work of Derek Bailey, Meredith Monk, Miya Masaoka, Karl Berger, and Joëlle Léandre; www.improvisedmusic.org/whoWeAre.html).[4]

What improvisational music and these musicians have in common is a shared commitment to solo and group improvisation as an art, a process that in many ways inverts what is "expected" by Western audiences. Rather than arriving at a gig to play previously conceived and rehearsed material, improvisers often arrive with no material and, depending on the context and agreement between participants, begin playing together without discussing aspects of musical form and function that are common to many performances (rhythm, pitch, tone, volume, etc.). This kind of performance requires a special kind of awareness, openness, and intentionality, working with others to create something in the moment without a net of composition, form, rhythm, melody, etc. It does not mean, however, that such performances are necessarily atonal or somehow do not use a musician's available technique and prowess. Quite the opposite, improvisation requires great skill, deep familiarity with the sounds and limitations of one's instrument (so that they may be overcome), and what some musicians call "big ears," the ability to hear possibilities, trajectories, and events as, if not before, they occur. Further, unlike improvisational theater, musical improvisations need not agree or be congruent. For example, one moment a musician can play in opposition to her bandmates, in an overpoweringly loud fashion, only to find that the rest of the band has dropped out and she has become a momentary soloist, to be joined moments later by someone playing in a different time signature quietly as she continues to play at top volume.

Finally, it is important to note that although many forms of jazz have much in common with Western music theory, this is not necessarily the case. This can be seen in Yusef Lateef's (1981) *Repository of Scales and Melodic Patterns*, Dizzy Gillespie's moves to study Afro-Caribbean (particularly Afro-Cuban) music, and John Coltrane's Africa suite. In other words, although straight ahead jazz has not only African roots but is in many ways driven as much by the exploration and contribution of non-Western music. Yet, jazz as a metaphor is nearly always a soloist model of improvisation and conceptualized as momentary movements within a structure.

Focusing on Laudan Nooshin's (2003) critical examination of encounters between Iranian (Persian) classical music (*musiq-e assil*) and Westerners' notions of improvising, the following section documents one example of how Western constructions of what improvising means have been used to marginalize non-Western musical traditions. Here, Nooshin raises points similar to those made by Lewis (1995, 2008) about the ways in which mainstream notions of jazz/improvising in many ways continues to silence non-mainstream constructions of jazz and improvised music in the West. The differences here lie in the ways in which such Western musical notions serve not only to marginalize particular non-Western musical ways of knowing but also in how they simultaneously served to colonize long-standing high status musical traditions—an explicit downgrading and marginalizing of a non-Western high status art music parallel to the Western canon (e.g., Beethoven, Bach, Hyden, Mahler, etc.).

Improvising in Musiqi-e Assil

Unlike the Western notions of improvising presented above, and similar non-Western constructions of improvising such as those associated with the Hindustani (N. Indian "classical") tradition, *musiq-e assil* involves placing largely preset musical phrases together in a spontaneously responsive fashion within relatively strict parameters. In order to be understood as well-conceived and tastefully executed, these improvisational phrases must be strung together in an aesthetically correct time and manner, constructions that are tightly bound and monitored according to rather particularized high status musical norms and values.

Nooshin's (2002) discussion of the role of improvisation in *musiq-e assil* is particularly salient to the points I wish to make here. Nooshin writes that, where

> *musiqi-e assil* literally means "pure music" or "noble music", and is also known as *musiqi-e sonnati* ("traditional music"). It is generally referred to outside Iran as Persian (or Iraninan) classical music, but I prefer to use local terminology, particularly since in Iran the term "classical" (*klasik*) denotes Western classical music.
>
> (p. 245, ft 8)

Along similar lines, she prefers the term Iranian music to Persian because "over recent decades the music has broadened in scope and has come to be widely regarded as a national music. . . . In Iran this, this music is described as Irani ('Iranian') and I follow this here" (Nooshin, 2002).

According to Nooshin (2002), discussions of what counts as composition and improvisation "continue to depend on the reification of particular aspects of music-making" (p. 242) that privilege composition over improvisation, the written over the performed, and, in so doing, Western music over "other" music. Traditionally, before the turn of the last century and ever-increasing contact with the West, *musiqi-e assil* did not involve any written composition. Performances were comprised solely placing together phrases and ideas in real time so that a composition formed as it was performed. Today, while this aspect of the tradition is alive and well, it is also not unusual, if not common, for music to be composed as a written score in Western musical notation. In order to better represent both the nature of this music and its relationship to improvisation, I quote Nooshin here at length.

> To the extent that creative performance lies at the heart of this music, it is usually described as "improvised," both by musicians and in the literature. Yet, whilst the concept of creativity—*khalaqiat*—has a long history in *musiqi-e assil*, as in so many other musical traditions there existed no equivalent of the Western concept of improvisation in Iran prior to the twentieth century. Creativity in performance was simply an accepted part of a tradition in which no distinction was made between the roles of composer and performer. However, this creativity was understood to be firmly grounded in a lengthy and rigorous training involving the precise memorization of a canonic repertory known since the late nineteenth century as *radif* (literally "series"). For the first Europeans to encounter *musiqi-e assil*, performed apparently spontaneously and without notation, the music must indeed have seemed improvised, particularly since they are unlikely to have been aware of the underlying learnt reparatory.
>
> (p. 244)

Although the term "improvising" has come to be adopted by Iranian musicians to describe their performance of *musiqi-e assil*, a musical colonization that Nooshin clearly outlines in her work, what counts as improvising largely remains the placing together of such forms.

For Nooshin (2002), such a distinction is important for at least the following two reasons. First, a construction of *musiqi-e assil* as improvised rather than composed allowed Western musicologists to classify the tradition as not only different but less than Western (composed, texted) music. As implied, this differentiation was made possible due to a (false) understanding that rational, prescribed composition is of greater value than purportedly spontaneous improvisation—a still-prevalent yet debunked understanding of music in which improvising is just playing whatever comes to mind without regard to skill or understanding where writing and reading music takes serious musicianship and knowledge. Second, similar to other discourses of othering that utilized text as a means to distinguish between Western superiority in culture and society and grant non-Western traditions a subaltern status (e.g., Appadurai, 2001; Bhaba, 1990; Minh-Ha, 1989; Said, 1993; Spivak, 1988)—Western music is constructed as a high art that affirm a Western sociocultural superiority (e.g., Agawu, 1992; Born & Hesmondhalgh, 2000; Solie, 1993; Tenzer, 2000).

Nooshin's discussion of improvising therefore has the following implications for the points I wish to raise in this chapter. On one hand, her discussion of what improvising means is an important reminder. What may at first seem improvised can in fact be a performance of previously scripted and studied material strung together to fit local aesthetic norms and values about the construction of sounds-as-music. Although Nooshin uses similar terminology, a distinction here should be made between what Ornette Coleman refers to as "spontaneous composition" and *musiqi-e assil*. While both constructions share an understanding of improvising as the result of years of training, thinking, listening, and practice, *musiqi-e assil* involves an assemblage of specific building blocks of sound (think sound Legos) that are assembled in particular ways to form a performance. While spontaneous composition often does involve placing together particular musical tropes that are the result of years of study, practice, and performance, it does not necessarily follow a set of rules and practice, as does *musiqi-e assil*.

Second, Nooshin reminds us of the constructed nature of improvising and composition, terminology that, as is the case with all language, conveys particular sets of norms and values about that which is named. Although inroads have certainly been and continue to be made to discussions of music, composition, and improvising in less colonizing language, improvising continues in many circles to be presented as the lesser pair in a static binary of composition-improvisation in many discourses about music. In sum, it often remains the case that there is both an understanding of approaches to music that are presented as a composition-improvisation binary. In this way, noting a tradition or musician's improvisation similarly often continues to privilege

Western constructions of music and musicianship, in which improvisation functions as a maker of difference as deficit.

Although my purpose here is not to review the myriad musical traditions that have been categorized as either less sophisticated or evolved due to their construction as improvised rather than composed music, it is important to note that traditions that range from the kinds of improvisational practices in which the AACM engage to other hundreds of years' old traditions (e.g., Hindustani, Northern Indian "classical" and Carnatic, Southern Indian "classical" music) have all been constructed at various points in their histories as subordinate to Western art music, an understanding based in no small part on their characterization as improvised rather than composed (scripted). In order to see more current incarnations of this perspective, one only need to turn to the construct of "structural listening" advocated by Schoenberg and Adorno that is alive and well in various strains of musicology and music history (see, for example, Edström, 1997; Subotnik, 1996). It is only recently that this concept has been challenged, critiqued for its Western, modernist leanings; the ways in which a composition and text-based construction of musical importance continues to reify notions of text over sense and score over performance; and how Western constructions of music are raced and gendered (e.g., Dell'Antonio, 2004; Gopinath, 2009; Lewis, 2008; Solie, 1993; Subonik, 1996). Thus, improvising can be understood as an act of composition through performance as well as the music that emerges from that unscripted composition can also be comprised of prescribed sonic building blocks that are assembled according to relatively static prescribed patterns.

Improvising, then, can be constructed as existing on a continuum, from a fully open structure with few if any rules to a highly ritualized act of creation through performance using largely prescribed sounds and patterns for their construction (see Table 5.1).

One way Lewis (2008) and other AACM members think of improvising is in relation to composition. This understanding of collective improvising is similar to *musiqi-e assil* in that musical structures *emerge* during performance, regardless of their relationship to previously learned phrases.

Table 5.1 Continuum of Musical Improvising

Kind of Improvising	*Open/Collective*	*Straight Ahead*	*Musiqi-e assil*
Structure	Emergent	Referential	Emergent
Prescribed content (patterns, modes, scales, melodies)	No	Yes	Yes
Prescribed rules	No	Yes	Yes
Notated/texted	No	Yes	Traditional=No Contemporary=Yes/No

Although meanings also emerge through straight ahead improvising, improvising straight ahead moves within and around prescribed structures in such a way that the prescribed structure is always *referenced*—whether working within or outside the structure of a song, the improvising is always in relation to the song's structure, its composition. What can make this understanding somewhat difficult for Western ears to grasp is the kinds of complications that Nooshin (2003) and Lewis (2008) present. For example while *musiqi-e assil* was not traditionally texted, it now has a tradition of texted, pre-performance composition. Although still the expression of many years of study, the emergent meanings in moment-by-moment musical decisions during performance that were once considered compositional in their own right have since become to be described as "improvised," even by musicians within the tradition.

Similarly, while improvising in the AACM can indeed be thought of in relation to composition, there are also many instances in which the organization's members have participated in open improvisations, musical interactions that necessarily draw upon years of dedicated practice and exploration. George Lewis' (2006) liner notes to his trio album *Streaming* with Muhal Richard Abrams and Roscoe Mitchell address this complication eloquently and I quote him here at length:

> Muhal, Roscoe and I have worked together since 1971, recording such albums as my *Shadowgraph* (Black Saint, 1977), Roscoe's *Nonaah* (Nessa, 1978), and Muhal's *Spihuonesty* (Black Saint, 1980). The success of our 2003 Venice Biennale event, for which we presented an open improvisation of nearly one and one-half hours, provided at least part of the impetus for us to focus on our collaboration as a trio. This recording can be viewed as one result of that impetus. The pieces presented here certainly do not represent anything like a culmination of our decades of musical research, exploration, and critical inquiry. Rather, what we hear are particular moments in time, in which improvisative musical experience proceeds directly from manifestations of trust and openness, and from visions of new models of community.

These three traditions of improvising—straight ahead jazz, open/collective improvising, and *musiqi-e assil*—share several key characteristics that have strong implications for conceptualizing students as improvisers. For example, these traditions demonstrate that improvisation is not simply an instance of expression, a musical or non-musical action based on what the improviser has in mind. Despite its often-spontaneous nature, improvising requires a multitude of intentional choices that are predicated on knowledge of the contexts, content, and possibilities for its expression.[5]

The ideas and feelings that become the sounds musicians produce are often split-second critical decisions made about what sounds or feels correct according to the musician's personal taste, ideas, and ideals. Such decisions are also inherently creative, making or remaking sounds in particular ways. Musical improvisers must possess the technical ability to get around on their instruments to articulate their ideas as they flow from awareness to musical expression, an ability that is most often gained through years of practice. Additionally, improvising is almost always in relation to some kind of musical structure, whether that structure serves as the framework a musician works within or around or if that structure is emergent from musicians' collective or open improvising.

Musical improvising is therefore simultaneously affective and cognitive. As these traditions illustrate, regardless of whether an improviser utilizes preset phrases and patterns or purposefully eschews such conventions, improvising requires technical skill, moment-by-moment critical decisions, and creativity. An examination of improvising therefore requires an exploration of both the frameworks one operates within or ignores as well as the processes through and around which those frameworks are negotiated on a moment-to-moment basis.

Thus, musical improving can be understood as the critically creative and skillful negotiation and/or creation of soundworks for feeling, thought, interaction, and other such processes. How, then, is improvising utilized as a means to discuss sociocultural phenomenon such as discourse, daily interactions, or research methodology? Returning to the vignette offered at the start of this chapter, this next section contextualizes the academic and social frameworks operating in Mr. Gutierrez's fourth grade classroom, then reexamines students' interactions through lenses of improvisation. As the following section presents, these forms of improvising, too, fall along a similar continuum.

Students as Improvisers

Contemporary schooling in many ways reflects a social efficiency construction of schooling (Kliebard, 2004), an ends-means vision of schooling that tends to cast students as interpreters of their teachers' ideas and ideals rather than critically creative agents of their own expression and accumulation of knowledge (e.g., Valli & Chambliss, 2007). This orientation often leads to constructions of students as binaries of compliant/resistant, on-task/off-task, and good/bad, conceptions of students that tend to miss the nuanced nature of what students do in classrooms and promote an understanding of students as either good (compliant) or "management issues" (noncompliant). In addition, such a vision of schooling tends to further negatively impact students of color (Anderson, 2009; Valenzuela, 2005).

As this section articulates, students are improvisers. Grounded in student interactions presented in the opening vignette, this section draws upon the examples of musical improvising presented earlier in the chapter to note the ways in which conceptualizing students as improvisers provides a more open lens for understanding students and the classroom structures in which they learn. Even in classrooms such as Mr. Gutierrez's, where their role is highly circumscribed, students work within and around academic and social classroom structures. Understanding students as improvisers also has the potential to move practical discussions of what students do in classrooms away from binary constructions of students where there are only two choices, following the teacher's expectations or being out of compliance, and towards the multiplicity of possibilities reflected in empirical research about students in classrooms (e.g., Metz, 1978; Pace & Hemmings, 2006; Page, 1991; Varenne & McDermott, 1998).

Such a change is significant if not vital in conceptualizing schools as it is in everyday classroom practices for at least the following two reasons. First, schools and schooling has long had a default positioning that conflates correctness and goodliness, rightness with singularity, and paths to such correct, good, and righteous singularity as both linear and sequential. From the "line of inquiry" used as a stick to measure college faculty's readiness for tenure and promotion to the measurable goals and objectives teachers use to create ends-means lesson plans, much of contemporary education is predicated on this kind of binary conflation. By this I mean a double move of first rendering complex processes, events, ecologies, and trajectories into educational binaries, then taking those simplified, narrowed parameters of a binary (this or that) and conflating them with implicit secondary values—turning the complex simple and then redefining that simplicity so that the redefinition stands in for the original complexity while simultaneously often subverting that message.

These educational binaries are most often good/bad, success/failure that are mutually exclusive and codependent so that successes require failures and goodness requires explicit badness. It is in no small part this kind of educational philosophy that frames children of color, poor kids, LGBTQ+ kids, Indigenous kids, kids whose home language is not English, and other groups' differences as deficits that simultaneously contribute to self-fulfilling prophecies of their marginalized status (e.g., Anyon, 1981; Cooper, 1892; Gilbert, 2013; Grande, 2004; Martinez, 2010; Mayo, 2013; Rist, 1973; Varenne & McDermott, 1998; Woodson, 1933).

This leads to the second reason that understanding students as improvisers is important. It is not only that the combination of narrow, linear pathways towards singular positive/negative binaries reaffirms an impression not only that there are only two choices and that educational actors must be categorized as one or the other in order that the binary be continually remade. Good students and bad students, good teaching and bad teaching. It is also that such falsely narrow and straightforward characteristics can be measured. Adding insult to injury, there is great care taken to ensure that measurement is quantifiably egalitarian and generally transparent. Although this is less the case for large companies like Pearson in recent national spates of standardized testing in the United States, there nevertheless remains an understanding among critics and proponents alike that anything called learning and teaching can be measured and that various biases and tendencies can be controlled.

In other words, *both* those for standardized assessments *and* those critiquing them for lacking internal or external validity all *agree* that students and teachers can be divided into good/bad binaries, a conceptual move that works to overlook the inherent racist, classist, and sexist histories and trends in testing. Understanding all studenting as improvising is therefore significant in that it provides a complex means for conceptualizing and articulating the wide variety of everyday educational interactions without needing to either reduce them to positive/negative binaries or compare a group against another for their choices to be rendered sensible.

Opening Vignette Revisited: The Players and Their Interactions

There were seven central players whose actions were explicitly part of the opening vignette: Donald, Marcy, Kevin, James, Isabella, Josep, and Mr. Gutierrez. Josep and Mr. Gutierrez's actions were the focus of the vignette and therefore central to the discussion of students as improvisers in this section, but the remaining four students' interactions are also germane to this discussion. The implicit improvisations of their unnamed peers who constitute the majority of people in this classroom context are similarly integral to this conversation.

Mr. Gutierrez was conducting a lesson-as-usual in the opening vignette. It functioned according to an IRE-participation structure (Mehan, 1979)—a teacher-initiated, student-respondent, teacher-evaluated structure ubiquitous in schools and a strategy that keeps pedagogical control firmly in teachers' hands while simultaneously creating a context in which teachers dole out correctness—and his presentation of content closely reflected classroom texts. In the lesson being delivered during the vignette, Mr. Gutierrez was reading sentences from the *Teacher's Edition* that were an exact reproduction of students' *Reading and Writing Notebooks* and then writing those sentences on the board exactly as they appeared in the text.

The reproduction of page 67 of the students' *Reading and Writing Notebook* in front of Mr. Guiterrez was a note to teachers about students' answers in bolded fuchsia: "Answers will vary. Possible answers are shown." The blank in sentence number three "Lorenzo woke at 5:00 A.M. and looked at the ________ street" is filled with the word "unlit." In keeping with his established classroom participation patterns, Mr. Gutierrez initiated an IRE sequence in which he solicited an answer from Donald, then wrote the sentence on the board, filling in the blank with Donald's answer, "quiet."

Although not the focus of the vignette, the following interactions also occurred while Josep crawled and Mr. Gutierrez continued with the lesson following an IRE participation structure. Marcy had her book open. But, instead of writing the correct answer on the page, she was writing a note on a piece of paper that she later passed to a friend. Isabella also had her workbook open with a paper on top and was drawing pictures, not bothering to look up when Mr. Gutierrez spoke or wrote on the board. In contrast, Kevin and James had their books open and were writing in answers to the blanks they hadn't filled in, holding a well-timed, whispered conversation as they did. When Mr. Gutierrez faced the board writing, the boys whispered writing the correct answers in their empty assignment page, when Mr. Gutierrez turned around to ask the next student the answer they had provided in the appropriate blank, both boys had their heads down looking at their book.

Then there is Josep. Josep managed to: exit his chair, crawl across the floor, stand up after making it across the classroom, and still remain unnoticed until he wished to be visible. To this point, Josep had to make increasingly greater noise thrashing around on the teacher's desk performing the task of "looking for something" until his movements were bordering on muppet-like and the sounds of his rustling objects about was louder than his teacher's voice before Mr. Gutierrez stopped the lesson to ask Josep what he was doing out of his seat and at the teacher's desk. Implicitly complicit in this task were not only his friends at whose desks he stopped to get their attention, letting them in on his crawl with a broad-look-at-this!-grin, but also the unmentioned peers, the majority of whom had seen Josep's crawl yet continued to interact as if nothing was out of the ordinary.

As noted in the introduction to this chapter, with the exception of Donald's reaction to Mr. Gutierrez, the remaining student interactions would most likely be categorized as off-task or non-educative behaviors in most discussions of classroom teaching practice. The difficulty with this categorization is that each of these students' behavioral choices are not dissimilar to Donald's, they are a direct response to the academic and social participation structures in Mr. Gutierrez's classrooms. Furthermore, students are building knowledge, learning about how to negotiate both academic content and expected interactional patterns. As the remainder of this section demonstrates, understanding students' interactions through differing conceptualizations of improvisation provide slightly different windows into the relationship between students' classroom interactions and the structures of classroom lessons.

Before continuing there is an important point I wish to note about Mr. Gutierrez's classroom in general and this lesson in specific. The ways in which he elected to conduct classroom lessons were not only the kinds of lessons his principal sought from teachers at the school but also reflected broader understandings of strong teaching throughout the district. Mr. Gutierrez was respected by his colleagues and his students' test scores, while not at the top of the district, were above district norms for the immigrant, second language learners who comprised the majority of the students in his class. He was also regularly observed by math and language arts "coaches" who not only saw the kinds of lessons presented in this chapter as appropriate but also worked to have teachers throughout the school conduct daily lessons in this fashion. In short, were it not for Josep's cross-classroom trek, there would have been little remarkable about the lesson described in the opening vignette, a pedagogical style lauded by his peers, administration, and district.

Students Improvising: Straight Ahead, Musiqi-e Assil, *and Collective Constructions*

Scholarship on musical improvising and that which uses musical improvisation to describe improvising in non-musical contexts demonstrates the importance of clarifying definitions of what improvising means, the structures in which one improvises, as well as who improvises. In order to demonstrate how students are improvisers, this section presents the opening vignette as three different musical improvisational models: straight ahead, *musiqi-e assil*, and collective improvising.

Improvising Straight Ahead

Because a straight ahead jazz model of improvising is comprised primarily of a soloist and accompanying group, this improvisational model tends to yield the impression that one person at a time has the lead and others are working to provide support or structure for the soloist. Similarly, because straight ahead improvising relies heavily on the melodic and rhythmic structure of a given tune for guidance, use of this model tends to operate from an understanding that player's improvising moves fluidly within a given structure in order for their actions to be perceived as consonant. From this conceptualization of improvising Josep was the most likely soloist as it was his interactions that are the focus of the vignette, despite, or, as I will suggest momentarily, *because of*, his teacher's firm grasp on the direction and content of the lesson.

From the locus of a straight ahead improvisational stance, Mr. Gutierrez's decisions about how to deliver the formal curriculum, the official knowledge students are intended to learn in classrooms (e.g., Apple, 2014; Kumashiro, 2008; Page, 1991), serves as the structures around and through which Josep improvised. Mr. Gutierrez elected to follow the score in a relatively literal fashion, delivering his interpretation of the *Teacher's Edition* with little to no variation on a theme. Mr. Gutierrez also chose to again follow an IRE participation structure to deliver this content. Student improvisations therefore could be seen as functioning within or around the content contained in their *Readinga nd Writing Workbooks* and an IRE participation structure.

Josep's improvising does just this. He completed the assignment; all the blanks in his workbook were filled in. Josep then waited until Mr. Gutierrez turned to the board to slip out of his chair and begin crawling across the floor, a maneuver he carried out with such skill that even I whose sole task was to observe the classroom did not at first notice him out of his seat. Josep timed his crawl so that he stopped each time his teacher turned to face the class and began crawling again as soon as Mr. Gutierrez turned back to the board to write another peer's answer. Josep's knowledge of these structures was such that not only was he not caught as he crawled, he was able to let his friends in on the act, pulling at pants legs and backpack straps to get their attention as he passed by them. Additionally, his improvisation changed in pitch and tone from purposefully unnoticed crawl to highly visible and increasingly loud rooting around Mr. Gutierrez's desk "for a pencil"—an interesting decision in light of the pencil that sat atop his open *Workbook* on the other side of the classroom.

Through this straight-ahead improvisational lens the remaining students and Mr. Gutierrez served as accompanists, keeping the tune moving forward and making the necessary adjustments to further enable Josep's improvising. Where Mr. Gutierrez kept lessons forward in terms of both academic content and participation structure, Josep's peers' interactions function on two levels simultaneously. At one level, they were performing their prescribed roles: responding to their teacher's questions, writing or correcting their answers, and giving Mr. Gutierrez the conversational space to lead the lesson.

On another level, their surface participation belied their covert subversion of those structures. It is precisely their appearance of compliance that enables Josep's actions. By the time Josep stood up and started rifling through the layer of papers and books on Mr. Gutierrez's desk, the majority of the class had become aware of his crawl, an awareness that spread like a silent wave throughout the room. Had any one student so much as intimated that something was afoot with Josep, Josep would have in all likelihood been busted. There is little room to further hide once one is on the floor on all fours between table groupings.

A straight ahead improvisational lens also lends itself to questions about the reasons for participants' interactions. Due to teachers' classroom roles, in order to understand the structures in classrooms necessitates an exploration of how teachers construct their relationship to the academic and social aspects of classrooms. As this chapter has already addressed Mr. Gutierrez's perspectives about academic content and documented his approach to classroom participation, I will instead turn to Josep's talk about his decision to crawl across the room.

Despite how it may seem from the vignette, Josep was a very strong student academically. He rarely struggled with his work, was often one of the first students finished, and on the district's 4-point grading scale regularly received "4s" in nearly all subjects on each report card. Yet, here is how Josep responded during an interview session to a question about his thoughts on lessons in Mr. Gutierrez's classroom:

> Yeah, you know, I'm bored in class most of the time. It's not that I'm really bad, it's just there's nothing for me to do, you know? I mean, I already get it and all so why go over it again?

In addition to academic content, it would appear as though one thing that Josep was learning is not only that he can "already get" the correct answers, he has also learned that classrooms are highly repetitive and structured places. He has also learned that he can use his knowledge about the flow of classroom lessons to get around them in ways that are much more exciting than participating in the lesson according to its prescribed structures.

WG: So, class was a bit exciting today.
Josep: Yeah (giggling).
WG: What were you looking for on Mr. Gutierrez's desk?
Josep: Aww, not much. Just, that, well, did you see what I did?
WG: Something other than looking on his desk (joking tone)?
Josep: Yeah, you know. I crawled across the room. I was just sitting there, listening to all the answers, again, after we filled 'em out and all, and I thought, hey, I bet I can crawl across the room, and I did!

Crawling across the room was not only a chance for Josep to get some rarely received movement. It was also an opportunity for him to critically and creatively express himself, as opposed to the boredom he often expressed when we talked, and use his knowledge of academic content and classroom participation structures in ways outside their relatively narrow confines. From this perspective, Josep's choice to crawl across the classroom floor can be seen as his intentionally enacting a creative solution to his boredom that also demonstrated his dissatisfaction with the content and form of classroom lessons. As can be seen in his talk above, this crawl was also an opportunity for Josep to explore an idea that sparked his interest, an opportunity that he felt he was largely denied due to the relatively rigid predictability and prescribed nature of daily lessons.

Unlike most understandings of classroom interactions where Josep's choices would be seen as counter to and removed from the task at hand, a straight ahead improvisational lens demonstrates the connection between the classroom structures Mr. Gutierrez enacted and Josep's negotiation of those structures. Josep had learned important yet unintended lessons about being a student in Mr. Gutierrez's classroom. He had come to understand that academic content was boring and classroom lessons were repetitive. Josep had also come to understand his role within these spaces as passive. Regardless of his particular interest or lack of interest in a given subject, the lessons moved apace without much input from him or any of his peers. Furthermore, in this single event, Josep had come to know that provided that he completed his work, he would be largely out from under his teacher's watchful eye and could exploit classroom interactional patterns to his and his peer's amusement.

A straight ahead model of improvising therefore provides a window into the academic and social structures of classrooms and how students move within or around these largely teacher-directed frameworks. It presents the possibility that students, as well as teachers, can serve as a soloist and how students explicitly and implicitly react to one another, enabling and constraining their improvising.

However, what a straight ahead lens has some difficulty dealing with is the multiplicity of soloist-possibilities at any given moment in time. For example, when Mr. Gutierrez was teaching and Josep was improvising, was Mr. Gutierrez or Josep the soloist—who was accompanying whom? What to make of Marcy, Isabella, Kevin, or James' interactions? Were they also soloists improvising as Josep did or were they accompanists? Similarly, what of the majority of students in the classroom whose improvising consisted largely of performing their roles in an uninterrupted fashion in order to not draw attention to the explicitly improvisational nature of Josep's crawl? Examining students' interactions through collective and Persian improvisational lenses are two answers to such questions.

In addition to an analysis of student interactions from a particular conceptual framework, all possible improvisational lenses necessitate an examination of the contexts and content of the structures around which students improvise as well as the intent behind these critically creative expressions. Because all of the examples in this chapter draw upon the opening vignette, rather than review the players' talk about their interactions or review the contexts that inform those interactions, I have elected to work in an additive fashion, adding points of analysis and referencing those explicated in previous models.

Student Improvising as Musiqi-e Assil

Improvising in *musiqi-e assil* is performed according to intricate sets of prescribed patterns and possibilities that participants weave together in response to one another's choices and an equally learned sense of appropriate expression of a given musical composition. The net result of such improvising is that while musicians are provided a multitude of musical possibilities within compositional structures and have hundreds of improvisational phrases at their fingertips, the implementation of those phrases is generally to be accomplished in accordance within highly particularized aesthetic notions of musical propriety.

Musicians are faced with the task of improvising so that they push at musical conventions to the utmost of their critically creative powers without crossing the boundaries of taste, form, and function. They must make decisions that are not so elementary as to not demonstrate their deep skill and knowledge yet do not stray outside of what is considered to be proper improvisational expression.

As with a straight ahead improvisational lens, examining students' classroom interactions from a *musiqi-e assil* perspective requires an exploration of the contexts and content of classroom academic and social structures. However, this understanding varies from a straight ahead jazz construction of improvising in three key ways.

The first key difference from a conventional jazz construction of improvising lies in an understanding of the student's relationship to classroom structures. An explication of student interactions through a *musiqi-e assil* lens constructs those interactions as accumulations of prescribed improvisational possibilities. Consider, for example, all student interactions other than Josep's crawling in the opening vignette. Marcy and Isabella utilize the same improvisational trope. Both girls have elected to place a piece of paper on top of their *Reading and Writing Workbooks* so that it is not immediately clear that either student is not working on the assignment. They have also both chosen to lean their bodies so that exactly what they are writing is not in their teacher's direct line of sight.

Kevin and James utilize these two improvisational tropes but in a slightly different manner. They too subtly lean towards one another so that they do not sway back and forth as Mr. Gutierrez turns to and from the board, lessening the chance that their talk will be noticed due to their physical actions. Where Marcy and Isabella place a paper on top of their worksheets in order to maintain the illusion that they are engaged in the lesson at hand, Kevin and James time their whispering so that it occurs largely while Mr. Gutierrez is facing the board. The nature of Kevin and James' improvising is also different from Marcy and Isabella's. Marcy and Isabella's improvising is enacted in a solo fashion; Kevin and James collaborate. Similarly, Kevin and James' improvising is verbal where Marcy and Isabella's improvising is non-verbal.

Although more complex in form and function, when seen using a *musiqi-e assil* understanding of music, Josep's interactions are a similar compilation of tropes. Like Kevin and James, his improvising is collaborative in nature and timed to Mr. Gutierrez's movements at the board. Without the implicit help of his peers, his crawl would have been immediately terminated and he would have faced the consequences of such an explicitly non-lesson-related set of choices. On the other hand, Josep's improvising is also a non-verbal expression of his ideas. To these tropes, Josep adds a depth of physicality often reserved for other non-teacher sanctioned interactions such as getting up to sharpen one's pencil or the use of a trip to the bathroom to walk up and down the halls. He also adds a non-verbal communication trope. As Josep crawls from his seat to the teacher's desk at the other side of the classroom, he stops to get his friends' attention and receive greater recognition of his actions from most his classmates in the process. This is an extension of a trope familiar to teachers that is often observed out of the corner of our eyes or heard as we move about, students wildly gesturing to get one another's attention or mouthing words to one another across the classroom. When he arrives at the teacher's desk, Josep applies these tropes in reverse, making progressively more noise without timing his movements to Mr. Gutierrez's dance at the board until he is noticed.

From a *musiqi-e assil* improvisational perspective, what makes Josep's improvising so impressive is not the uniqueness of the tropes he uses but the ways in which he combines them. It is the specific way in which Josep layers these tropes that enables him to successfully crawl from one end of the classroom to the other.

While the student's improvising clearly demonstrates the depth of his knowledge about the academic and social structures of lessons in Mr. Gutierrez's classrooms, all improvising functions *within* these structures. At no point does any student improvisation break the flow of the lesson. Instead, maintaining these surface structures is integral to executing improvisational tropes that are implicitly designed so that they operate in the classroom underlife—interactions that lie outside formal lesson structures and content but do not make waves across the lesson's surface (Gutierrez, Rymes, & Larson,1995; Page, 1991).

Additionally, improvising in *musiqi-e assil* constructs all student interactions as improvisations. Perhaps the most clear use of this understanding can be seen when this lens is applied to the majority of students in Mr. Gutierrez's room who enable Josep's crawl, as well as Marcy, Isabella, Kevin, and James' improvising, through their apparent inaction. Rather than understanding these students' interactions as non-improvisatory, a Carnatic improvisational perspective lends the possibility that they are enacting a particular improvisational trope: enacting classroom participation structures as expected as if nothing peculiar is happening. As with the other tropes presented in this section, this trope is most likely familiar to many teachers. In my own teaching career, I remember a particular instance where all 33 fifth graders in my classroom chose, other than an occasional giggle, to act as if there was nothing out of place instead of letting me know about a particularly large piece of lettuce stuck to the outside of my beard.

A *musiqi-e assil*-type improvisational perspective has one further advantage. From this framework of improvising, all classroom interactions can be considered improvisatory. In practical terms, this means that not only can all students' interactions be considered improvisations but also their teacher's interactions as improvisational. What makes the consideration of all classroom interactions as improvisational particularly useful in considering educational contexts is that this construction not only illuminates classroom structures and classroom actors' decisions in relation to these structures, but it also facilitates comparison between the tropes utilized by students and teachers.

For example, Mr. Gutierrez was in all likelihood enacting a similar trope to the ones used by students who acted as if there was nothing out of the ordinary as Josep crawled across the floor. He may well have noticed Marcy and Isabella's drawing or heard Kevin and James' whispering and elected not to negatively sanction their improvisations because they did not break the ebb and flow of the lesson he was conducting. Thus, through the lens of improvising in the traditions of Carnatic music and *musiqi-e assil*, all student interactions are considered to be improvisational, an understanding that provides the possibility that all classroom interactions, student and teacher alike, are improvisations.

Collectively Improvising: Collaboratively Emergent Meanings

Understanding students' classroom interactions as a collective improvisation functions in many ways in the inverse of either a straight ahead or a Carnatic/*musiqi-e assil* improvisational lens. Visions of collective improvising share an understanding that musical meanings can be emergent rather than prescribed and that a polyphony of improvisational decisions need not be in concert in order to be meaningful (e.g., Lewis, 2008).

When applied to classroom interactions, this construction is in many ways similar to improvising in *musiqi-e assil*. For example, if understood as collective improvising, all classroom actors' interactions are improvisations that work in concert with one another in some fashion, chains of intent and expression, cause and effect that are inexorably interrelated. As is the case with straight ahead jazz and *musiqi-e assil*, collective improvising is also predicated on the understanding that improvising requires knowledge and skill of both classroom structures and the pathways through them. Where a collective improvisational lens differs from both straight ahead and *musiqi-e assil* improvisational models is in how meanings are constructed, the interrelated nature of those constructions, and the openness of interactional possibilities in classroom interactions.

In musical collective improvisations, any player at any time can take the lead and each player's decisions impact all other improvisational choices. There is, however, often no specific lead but instead one or more voices that emerge as dominant at any one time, voices that can as quickly be removed of their centrality to the group's improvising. As there is no need for musical consensus, a player can elect to follow or disregard another player's improvisational decision. Because there is no preset framework and each person's decision necessarily impacts another's, players must especially attend to others' choices as they enact their own.

This collaboratively engaged understanding of improvising has particular implications for understanding how classrooms function. Regardless of any one person's intent or expression of that intent, classroom meanings are not prescribed but emergent. In the vignette this can be seen in the differences between the meanings Mr. Gutierrez's lesson appears designed to impart and the unintended ideas about school and content that Josep learned during the course of those lessons.

Such meanings are not mutually exclusive. From a collectively improvised perspective, Mr. Gutierrez is delivering the content he intends and Josep is learning content that his teacher does and does not intend for him to come to know. In addition, because all student interactions can be understood as improvisations, the weight of any one person's interactions is not necessarily less meaningful than any other person's at any given moment in time. For example, without the majority of students' apparent participation in the lesson, Josep might not have been able to leave his seat, let alone spend the better part of three minutes crawling across the classroom floor.

Further, because meanings are layered and emergent, the meanings of classroom lessons, the multiplicity of what teachers teach and students learn, is necessarily complex and cannot be easily reduced into static binaries. Consider Marcy's note writing while Josep crawled across the classroom floor. This can be seen as learning about how to work within classroom structures or as practice in writing notes. In a more traditional construction of student interactions, this particular expression of her learning would most likely be considered as lying outside the boundaries of the lesson at hand. However, from a collectively improvised perspective, her choices are directly related to the lesson. In her note writing, Marcy is observing classroom participation structures, allowing the conversational space for responses to Mr. Gutierrez's initiations, and, as her note is certain to contain at least an adjective or two, she is also practicing the concepts Mr. Gutierrez is instructing.

Finally, a collective improvisational lens affords the user an opportunity to place all these meanings side-by-side in order to examine the emergent constructions of knowledge as they are enacted in classrooms rather than in relation to a prescribed framework. This ability does not preclude the possibility that there is indeed a dominant, teacher-directed set of academic and social structures in classrooms. What a lens of collective improvisation affords the user is the possibility that this structure, while in many ways the driving force in the classroom, is not the only or always the dominant structure operating during any moment of a classroom lesson. For example, it could be argued that while Josep was crawling across the floor, his particular improvisational choices recast classroom structures to the point that his crawling was central and Mr. Gutierrez's relatively steadfast enactment of the Teacher's Edition served primarily in support of Josep's decisions.

Improvising, Race, and Culture

Interpretive studies of classrooms have long documented how students of color and poor students, in spite of the rich understandings and traditions they bring to educational contexts (e.g., Yosso, 2005), are often marginalized by mainstream constructions of knowledge in schools (e.g., Anderson, 2009; Anyon, 1981; Rist, 1970; Varenne & McDermott, 1998; Woodson, 1933). Similar understandings can and often do lead to schooling in which traditionally marginalized students' status is reified by their teachers approach to classroom management and teacher expectations (e.g., Alonso, Anderson, Su, & Theoharis, 2009; Cooper, 1892; Foley, 2010; Willis, 1977). For example, as Foley (2010) notes, parallel to other discussions of agency and authority in classrooms (e.g., Ortner, 2006; Pace & Hemmings, 2006), different students have different degrees of wiggle room to negotiate in, through, and around classroom interactional patterns and teacher expectations, what I call classroom scores. Further complicating matters, although there is certainly a greater tendency for Anglo teachers to regard students of color's interactions as outside of implicitly mainstream classroom norms and values (e.g., Heath,

2006; Valencia, 2010), teachers of color can also hold students like themselves in low regard (e.g., Prime & Miranda, 2006).

Such was the case in Mr. Gutierrez's classroom and at the school in which he taught. For example, although Armenian students were as likely to be first generation, non-"native" English speakers as their Latino/a peers, no Armenian students were considered to be English Language Learners in the four classrooms I observed for this study. This is no small matter particularly in light of the fact that I needed to send explanatory letters home about this project in English, Spanish, and Armenian in order that guardians could be fully informed about the project. Additionally, the school district categorized Armenian students as "White, non-Hispanic."

These factors combined to create a context in which Armenian students often led lives that were in many ways closer in experience to their similarly first generation Latino/a peers but held all the sociocultural markers of Whiteness.

Additionally, Mr. Gutierrez spoke to me on more than one occasion of how "the Armenian kids tend to get it but the Latino/a kids often need more help." While there is not the space to delve into this more fully here, Mr. Gutierrez was himself a second generation Latino who grew up in a middle class home and spoke English at home, learning Spanish later in school.

Moreover, Josep was an outstanding student by all academic markers. He received the numerical equivalent of A's in all subject areas (scores of 4 on a four-point scale) and had test scores that were well above the district average and in the highest quarter of scores in the district. As a result, Josep did have more wiggle room than many of his peers. For example, before crawling across the room, Josep had answered all the fill-in-the-blank questions correctly, leaving Mr. Gutierrez little recourse other than to insist that he return to his seat. Yet, in spite of his status, Josep was also a student in trouble, a student who Mr. Gutierrez kept a close eye on throughout daily lessons and who was occasionally blamed for things he did not do as a result of that consistent attention from his teacher.

Therefore, from one perspective, Josep may indeed have had more wiggle room than his peers to enact his epic classroom crawl. However, this wiggle room does not account for either the extreme skill and knowledge with which Josep improvised or his peers' improvisations described above. Of similar importance, neither Marcy or Isabella's first generation Latina status nor Isabella's low socioeconomic status kept either student from improvising. It is in these ways that an understanding of students as improvisers simultaneously provides a more robust means for conceptualizing studenting and creates the opportunity for considering important differences in students' relations to sociocultural norms and values.

The overarching point here is that there are multiple understandings of what improvising means and how that improvising functions in practice. Similar to the three examples provided here, they run the gamut from open forms of critical creation to assembling possibilities in the moment according to prescribed procedural patterns and rules. How one is, their way of being, is ontogenic and epistemogenic, ever-unfolding processes of being and knowing in all their forms. From this perspective, regardless of the improvisational metaphor used to view student interactions, all student interactions can be understood as improvisational. Additionally, as presented in this section, the consideration of students' classroom interactions as improvisational presents several important advantages to understanding what students and teachers do in classrooms. The following brief concluding section encapsulates these possible advantages as well as suggests other aspects of education to which an improvisational lens might prove fruitful.

Conclusion: Students, Improvising, and the Complexity of Classroom Interactions

Students' classroom interactions are complex, interrelated, and concurrent. As I have documented here, conceptualizing students' interactions as improvisations can help move those concerned with what happens in classrooms away from prescribed, structural notions of teaching and learning and towards understanding students' classroom interactions as nuanced and multifaceted. This is not merely a semantic difference but a conceptual shift in understanding how to make sense (Ellsworth, 2005) of educational contexts and the ideas and ideals that inform those spaces, places, and identities.

Understanding students as improvisers is not trivial. In our particularly standardized times, where the formal curriculum is often a large part of a classrooms' score (Anderson, 2009; Valli & Chambliss, 2007; Wills, 2006), students and their teachers are consistently measured by students' scores on standardized assessments given in accordance with both state and federal policy mandates. These assessments equate student knowledge with test scores. The difficulty with this construction is not that it does not measure some of what students know. Rather, the problem lies in that students, and by extension their teachers, are reduced to their responses on prescribed annual measurable goals and objectives, often referred to as "benchmarks" or "grade level/appropriate standards." Additionally, while the standards themselves are often expressed in rather general terms, their expression on annual standardized assessments is most often in the form of known answer questions. Missing from this equation is the majority of what students do in classrooms on a moment-to-moment basis, the academic content and social expectations that frame those interactions, and how teachers and students co-construct these meanings together.

Measuring students according to preestablished criteria tends to regard ideas and ideals outside the scope of those specific goals and objectives as inconsequential. As a result, both intended and unintended classroom lessons that are not prescribed, as well as the purposeful and inadvertent consequences of those lessons, are often overlooked or dismissed. The problem with this perspective is that the majority of what students do in their daily lives as students is comprised of just these kinds of lessons. In short, setting learning that lies outside of measurable goals and objectives renders insignificant much of what students come to know in classrooms, both in terms of daily lessons and on standardized assessments. Furthermore, the nested layers of trickle-down of measurable goals and objectives—from federal policies to teacher's lesson plans—create a feedback loop that tends to preclude student inquiry and input into daily lessons.

Conceptualizing student interactions as improvisations is one solution that addresses such concerns. It should be noted that the boundaries between lenses of student improvising I outline here are indeed fluid. My rationale for providing these different improvisational possibilities was not to suggest that one is more powerful than another but instead to demonstrate the viability and versatility of these lenses for conceptualizing what students do in classrooms.

Furthermore, threaded throughout discussions of the various forms of improvising presented above is an understanding of improvising as an intentionally critical act, a notion of improvising that I wish to make explicit here. My use of the terms critical and criticality are aligned not only with critical improvisation studies but also with such fields as critical feminism (e.g., Bernal, Elenes, Godinez, & Villenas, 2006; Lather, 1991), Critical Race Theory (e.g., Ladson-Billings & Tate, 1995; Yosso, 2005), and queer theory (e.g., Pinar, 1998; Quinn & Meiners, 2010). While there are certainly tensions in and between these fields as to what "critical" means and to what ideas and ideals should be foregrounded in such studies, there is a general understanding in which criticality enunciates an attention to lines of power (sociocultural, socioeconomic, sociopolitical, etc.) and their often negative impact on those with the least amount of available power, social, political, economic, or otherwise. Additionally, it also bears mention that, as in the case of so-called classical improvisational music such as *musiqi-e assil* and some understandings of straight ahead jazz, there are forms of improvisation in which the purpose is to intentionally utilize established norms and values.

However, within a Western musical context, as is the case in American forms of jazz and improvising as practiced by members of the AACM for example, there is a history of improvising as speaking against established musical norms and values—ideas and ideals that similarly reflect broader sociocultural norms and values. As Lewis (2008) details, members of the AACM's use of improvising was not only compositional but also part of a broader and continuing movement towards sociocultural, sociopolitical, socioeconomic, and musical liberation and liberation for African Americans. Along similar lines, there is a long history in American jazz of choices about the kinds and length of improvising as forms of social activism.

From this perspective, conceptualizing students as improvisers maintains important understandings of students' agency (for a strong discussion of agency, Ortner, 2006)—their ability to use their available resources to negotiate explicit and implicit structures that tend to be imposed by those with greater classroom authority in schools, teachers and other adult school actors. Similarly, understanding all student interactions as improvisational creates the theoretical space for even apparent student inaction to serve as an implicit (or explicit) critique of dominant classroom norms and values, as was the case for the many students who noticed Josep's crawl but carried on as if nothing was out of the ordinary.

In these ways, constructing students as improvisers encompasses rather than rejects an understanding of students, even very young students, as intentional social actors who can and do use their available resources to evade, subvert, interrupt, or reject academic and social structures. Similarly, examining how students improvise makes explicit classroom participation and other such structures that serve as the frameworks for students' improvising. Additionally, while not necessarily the focus of this paper, a focus on students' agency as intentionally critical and creative improvisers can provide a means to view traditionally marginalized students as working within, through, and according to meaningful ways of knowing and being—ways of making sense of a classroom environment that can lie outside mainstream constructions of how classrooms function and how students should interact within those frameworks. In so doing, understanding students as improvisers can provide an opportunity to see how students interrupt, subvert, or coopt dominant notions of teaching and learning similar to the focus on Josep's interactions in this chapter.

Yet student improvising can also serve to reify classroom structures as well as dominant norms and values. It is certainly the case that it is always important to question whom such structures, ideas, and ideals benefit and at whose expense such benefits exist. However, it is also important to note that some structures, such as those that provide space for all students to have a voice, and some dominant norms and values, such as those that insist on a lack of physical and verbal violence, would not necessarily be harmful to reproduce.

Additionally, as I have documented here, like its musical counterparts, improvising is a critically creative act that is simultaneously political in its relation to questions of agency and choice and aesthetic. In these ways, conceptualizing students as improvisers contributes to the growing conversation at the intersection of aesthetics/politics and the ways in which individuals' sense-making is theorized and enacted both in and out of education (Ellsworth, 2005; Geurts, 2002; Kim-Cohen, 2009; Howes, 2003; Roelofs, 2009; Springgay, 2008). Understanding students as improvisers provides the space for educators and educational researchers to consider students according to what is meaningful to students in a way that is related to yet independent from now taken-for-granted notions of what students are to do in classrooms.

In sum, students use their available resources to make myriad moment-by-moment interactional choices to work within, around, through, or against explicit and implicit academic and social classroom structures. As such, students are improvisers. As I have documented here, the spectrum of improvising provides those who wish to understand what students do on a daily basis in educational contexts a rich, fluid means for conceptualizing what studenting means and the contexts that inform those choices.

From this perspective, the question is neither whether or not students are improvisers nor what interactions are improvisational. Rather, the question is in what kinds of improvising are students engaged? Further, what ways of being and knowing can such improvising engender, how well do those trajectories provide students the opportunity to clearly and respectfully express themselves, and what can students' improvisations help us understand about the educational ecologies in which they negotiate being, meaning, and sense?

Notes

1 All names are pseudonyms. In the case of *Geared for Success*, the program referred to later in this chapter, I have elected to give the program a pseudonym but quote from the text without altering its content in order to provide the reader with as accurate a representation as possible.

2 See, for example, the special issue of Leonardo Music Journal (2010, v20 n1) on improvisation and George E. Lewis' and Benjamin Piekut's (2016) *The Oxford Handbook of Critical Improvisation.*

3 Although there is continuing and growing attention to the AACM in contemporary jazz in general and among working musicians specifically, their impact and importance continues to often be overshadowed by the canonical narrative that now includes John Coltrane, has brief mention of Ornette Coleman, and overlooks master musicians such as George E. Lewis, Muhal Richard Abrams, and Anthony Braxton, whose corpus lies outside of more mainstream jazz explorations (e.g., Jack DeJonette).

4 It is important to note that this list is by no means exhaustive and that there is an endless, ongoing list of important improvisers in music who approach the potential of improvising and the possibilities for its conceptualization and expression from hundreds of different perspectives. Part of the difficulty for making the case I seek to argue here is this stream of possible models and listeners' particular preferences or awareness of one form or another. Rather, my overarching point regards the potential for both continua of kinds of improvising and their expression to serve as models for reconsidering and possibly reconceptualizing students.

5 For a complex series of conversations and demonstrations about the relationships between composition, musical expression, rhythm, melody, harmony, and improvisation, see/hear master musician Steve Coleman's website: http://m-base.com, and his free members website: http://members.m-base.net.

References

Agawu, K. (1992). Representing African music. *Critical Inquiry, 18*, 245–266.

Alonso, G., Anderson, N., Su, C., & Theoharis, J. (2009). *Our schools suck: Students talk back to a segregated nation on the failures of urban education.* New York: NYU Press.

Anderson, K. S. (2009). *War or common cause? A critical ethnography of language, education policy, race, and cultural citizenship.* Charlotte, NC: Information Age.

Anyon, J. (1981). Social class and the hidden curriculum of work. In H. A. Giroux, A. N. Penna & W. F. Pinar (Eds.), *Curriculum and instruction: Alternatives in Education* (pp. 3–42). Berkeley, CA: McCutchan.

Apple, M. W. (2014). *Official knowledge: Democratic education in a conservative age* (3rd ed.). New York: R outledge.

Bailey, D. (1993). *Improvisation: Its nature and practice in music*. New York: DeCapo Press.

Berliner, P. F. (1994). *Thinking in jazz: The infinite art of improvisation*. Chicago: University of Chicago Press.

Bernal, D. D., Elenes, C. A., Godinez, F. E., & Villenas, S. (2006). *Chicana/Latina education in everyday life: Feminista perspectives on pedagogy and epistemology*. Albany, NY: State University of New York Press..

Borko, H., & Livingston, C. (1989). Cognition and improvisation: Differences in mathematics instruction by expert and novice teachers. *American Educational Research Journal, 26*(4), 473–498.

Born, G., & Hesmondhalgh, D. (Eds.). (2000). *Western music and its others: Difference, representation, and appropriation in music*. Berkeley, CA: University of California press.

Cazden, C. (1988). *Classroom discourse: The language of teaching and learning*. Portsmouth, NH: Heinemann.

Coleman, O. (1983, June). Prime time for harmolodics. *Down Beat*, 54–55.

Cooper, A. C. (1892). *A voice for the South by a Black woman from the South*. Xenia, OH: Aldine Printing House.

Dance, J. L., Gutierrez, R., & Hermes, M. (2010). More like jazz than classical: Reciprocal interactions among educational researchers and respondents. *Harvard Educational Review, 80*(3), 327–351.

Davis, B., & Sumara, D. (2006). *Complexity and education: Inquiries into learning, teaching and research*. Mahwah, NJ: Lawrence Erlbaum.

Diamondstone, J. V. (1999). Tactics of resistance in student-student interaction. *Linguistics and Education, 10*(1), 107–137.

Doll, W. E. Jr., Fleener, M. J., Trueit, D., & St. Julien, J. (Eds.). (2005). *Chaos, complexity, curriculum &culture: A conversation*. New York: Peter Lang.

Edström, O. (1997). Fr-a-g-me-n-ts: A discussion on the position of critical ethnomusicology in contemporary musicology. *Svensk Tidskrift for Musikforskning (Swedish Journal of Musicology), 79*(1), 9–68.

Ellsworth, E. A. (2005). *Places of learning: Media, architecture, pedagogy*. New York: RoutledgeFalmer.

Erickson, F. (1982). Classroom discourse as improvisation: Relationships between academic task structure and social participation structure in lessons. In L. C. Wilkinson (Ed.), *Communicating in the classroom* (pp. 153–181). New York: Academic Press.

Erickson, F. (2004). *Talk and social theory: Ecologies of speaking and learning in everyday life*. Malden, MA: Polity Press.

Fischlin, D., & Heble, A. (2004). *The other side of nowhere: Jazz, improvisation, and communities in dialogue*. Middletown, CT: Wesleyan University.

Flinders, D. T., Noddings, N., & Thornton, S. J. (1986). The null curriculum: Its theoretical basis and practical implications. *Curriculum Inquiry, 16*(1), 33–42.

Foley, D. E. (2010). *Learning capitalist culture: Deep in the heart of Tejas* (2nd ed.). Philadelphia: University of Pennsylvania Press.

Foster, M. (2001). Pay Leon, pay Leon, paleontologist: Using call-and-response to facilitate language mastery and literacy acquisition among African American students. In S. Lanehart (Ed.), *Sociocultural and historical perspectives on African American English* (pp. 247–266). Reading, MA: John Benjaminus.

Gershon, W. S. (2006). Collective improvisation: A theoretical lens for classroom observation. *Journal of Curriculum and Pedagogy, 3*(1), 104–136.

Gershon, W. S. (2011). Towards a sensual curriculum. *Journal of Curriculum Theorizing, 27*(2), 1–16.

Geurts, K. L. (2002). *Culture and the senses: Bodily ways of knowing in an African community*. Berkeley, CA: University of California Press.

Gilbert, A. (2013). *Courageous pedagogy: Enacting critical science education*. Charlotte, NC: Information Age Press.

Grande, S. (2004). *Red pedagogy: Native American social and political thought*. Lanham, MD: Rowman & Littlefield.

Groesbeck, R. (1999). Cultural constructions of improvisation in *Tayampaka*, a genre of temple instrumental music in Kerala, India. *Ethnomusicology, 43*(1), 1–32.

Gutierrez, K., Larson, J., & Rymes, B. (1995). Script, counterscript, and underlife in the classroom: James Brown versus Brown v. Board of Education. *Harvard Educational Review, 65*(3), 445–471.

Hallam, E., & Ingold, T. (2008). *Creativity and cultural improvisation*. Oxford, UK: Berg.

Hanley, M. S. (2007). Old school crossings: Hip-hop in teacher education and beyond. *New Directions for Adult and Continuing Education, 115*, 35–44.

Heath, S. B. (2006). *Ways with words: Language, life, and work in communities and classrooms* (2nd ed.). Cambridge, UK: University of Cambridge Press.

Howes, D. (2003). *Sensual relations: Engaging the senses in culture & socialtheory*. Ann Arbor, MI: University of Michigan Press.

Jackson, P.W. (1992). Conceptions of curriculum and curriculum specialists. In P. Jackson (Ed.), *Handbook of research on curriculum* (pp. 3–40). New York: Palgrave Macmillan.

Kamoche, K., Cunha, M. P., & Cunha, J. V. (2003). Towards a theory of organizational improvisation: Looking beyond the jazz metaphor. *Journal of Management Studies, 40*(8), 2023–2051.

Kim-Cohen, S. (2009). *In the blink of an ear: Towards a non-cochlear sonic arts*. New York: Continuum.

Kondo, D. K. (1990). *Crafting selves: Power, gender and discourses of identity in a Japanese workplace*. Chicago: University of Chicago Press.

Kumashiro, K. (2008). *The seduction of common sense: How the right has framed the* debate *on America's schools*. New York: Teachers College Press.

Ladson-Billings, G., & Tate, W. F. (1995). Toward a critical race theory of education. *Teachers College Record, 97*(1), 47–68.

Langhout, R. D. (2005). Acts of resistance: Student (in)visibility. *Culture & Psychology, 11*(2), 123–158. doi: 10.1177/1354067X05052348

Lateef, Y. (1981). *Repository of scales and melodic patterns*. New York: Fana Music.

Lather, P. (1991). *Getting smart: Feminist research and pedagogy within/in the postmodern*. New York: Routledge.

Lewis, G. E. (1996). Improvised music after 1950: Afrological and Eurological perspectives. *Black Music Research Journal, 16*(1), 91–122.

Lewis, G. E. (2008). *A power stronger than itself: The AACM and American experimental music*. Chicago: University of Chicago.

Lewis, G. E., & Piekut, L. (2016a). *The Oxford handbook of critical improvisation studies, volume 1*. Oxford, UK: Oxford University Press.

Lewis, G. E., & Piekut, L. (2016b). *The Oxford handbook of critical improvisation studies, volume 2*. Oxford, UK: Oxford University Press.

Lobman, C. (2006). Improvisation: An analytic tool for examining teacher child interactions in the early childhood classroom. *Early Childhood Research Quarterly, 21* (4), 455–470.

McCluskey, A. T., & Smith, E. M. (Eds.). (2001). *Mary McLeod Bethune: Building a better world, essays and selected documents*. Bloomington, IN: Indiana University Press.

Mayo, C. (2013). *LGBTQ youth and education: Policies and practices*. New York: Teachers College Press.

Mehan, H. (1979). *Learning lessons: Social organization in the classroom*. Cambridge, MA: Harvard University Press.

Metz, M. H. (1978). *Classrooms and corridors: The crisis of authority in desegregated secondary schools*. Berkeley, CA: University of California Press.

Meyer, A., Frost, P. J., & Weick, K. C. (1998). The organization science jazz festival: Improvisation as a metaphor for organizing. {Special Edition] *Organization Science, 9*(5).

Mirón, L. F., & Lauria, M. (1998). Student voice as agency: Resistance and accommodation in inner-city schools. *Anthropology & Education Quarterly, 29*(2), 189–213.

Mitchell, R. (2010). Cultural aesthetics and teacher improvisation: An epistemology of providing culturally responsive service by African American professors. *Urban Education, 45*(5), 604–629. doi: 10.1177/0042085909347839

Monson, I. (1997). *Saying something: Jazz improvisation and interaction*. Chicago, IL: University of Chicago Press.

Montuori, L. (2003). The complexity of improvisation and the improvisation of complexity: Social science, art and creativity. *Human Relations, 56*(2), 237–255.

Nooshin, L. (2003). Improvisation as "other": Creativity, knowledge and power—the case of Iranian classical music. *Journal of the Royal Musical Association, 128*, 242–296.

Ortner, S. B. (2006). *Anthropology and social theory: Culture, power, and the acting subject*. Durham, NC: Duke University Press.

Pace, J. L., & Hemmings, A. (Eds.). (2006). *Classroom authority: Theory, research, and practice*. Mahwah, NJ: Lawrence Erlbaum.

Page, R. N. (1990). Games of chance: The lower-track curriculum in a college-preparatory high school. *Curriculum Inquiry, 20*(3), 240–281.

Page, R. N. (1991). *Lower-track classrooms: A curricular and cultural perspective*. New York: Teachers College Press.

Pinar, W. F. (Ed.). (1998). *Queer theory in education*. New York: Routledge.

Pinar, W. F., & Irwin, R. L. (2005). *Curriculum in a new key: The collected works of Ted Aoki*. Mahwah, NJ: Lawrence Erlbaum.

Prime, G. M., & Miranda, R. J. (2006). Urban public high school teachers' beliefs about science learner characteristics implications for curriculum. *Urban Education, 41*(5), 506–532.

Quinn, T., & Meiners, E. R. (2010). *Flaunt it! Queers organizing for public education and justice*. New York: Peter Lang.

Rist, R. C. (1970). Student social class and teacher expectations: The self-fulfilling prophecy in ghetto education. *Harvard Educational Review, 40*(3), 411–451.

Roelofs, M. (Ed.). (2009). Aesthetics and race: New philosophical perspectives. *Contemporary Aesthetics*, Special Volume 2. Retrieved from www.contempaesthetics.org/new volume/pages/journal.php?volume=41

Sawyer, R. K. (2001). *Creating conversations: Improvisations in everyday discourse*. Catskill, NJ: Haper Press.

Sawyer, R. K. (2004). Creative teaching: Collaborative discussion as disciplined improvisation. *Educational Researcher, 33*(2), 12–20.

Sharma, A. (2008). Portrait of a science teacher as a bricoleur: A case study from India. *Cultural Studies of Science, 3*, 811–841.

Shumate, E. D., & Wills, H. P. (2010). Classroom-based functional analysis and intervention for disruptive and off-task behaviors. *Education and Treatment of Children, 33*(1), 23–48.

Solie, R. (1993). *Musicology and Difference: Gender and sexuality in musical scholarship*. Los Angeles: University of California Press.

Spindler, G., & Hammond, L. (2006). *Innovations in educational ethnography: Theory, methods and results*. Mahwah, NJ: Lawrence Erlbaum.

Springgay, S. (2008). *Body knowledge and curriculum: Pedagogies of touch in youth and visual culture*. New York: Peter Lang.

Tenzer, M. (2000). *Gamelan gong kebyar: The Art of twentieth-century Balinese music*. Chicago: University of Chicago Press.

Thornborrow, J. (2003). The organization of primary school children's on-task and off-task talk in a small group setting. *Research on Language and Social Interaction*, *36*(1), 7–32.

Thorne, G. (1993). *Gender play: Girls and boys in school*. Rutgers, NJ: Rutgers University Press.

Tuck, E., & Yang, K. W. (Eds.). (2014). *Youth resistance research and theories of change*. New York: Routledge.

Valencia, R. R. (2010). *Dismantling contemporary deficit thinking: Educational thought and practice*. New York: Routledge.

Valenzuela, A. (Ed.). (2005). *Leaving children behind: How "Texas-style" accountability fails Latino youth*. New York: State University of New York Press.

Valenzuela, A., Prieto, L., & Hamilton, M. P. (Ed.). (2007) No child left behind (NCLB) and minority youth: What the qualitative evidence suggests. [Special Edition]. *Anthropology and Education Quarterly*, *38*(1).

Valli, L., & Chambliss, M. (2007). Creating classroom cultures: One teacher, two lessons, and a high-stakes test. *Anthropology Education Quarterly*, *38*(1), 57–75.

Varenne, H., & McDermott, R. (1998). *Successful failure: The schools America builds*. Boulder, CO: Westview P ress.

Willis, P. (1977). *Learning to labor: How working class kids get working class jobs*. New York: Columbia University Press.

Wills, J. S. (2006). Authority, culture, context: Controlling the production of historical knowledge in elementary classrooms. In J. L. Pace & A. Hemmings (Eds.), *Classroom authority: Theory, research, and practice* (pp. 33–62). Mahwah, NJ: Lawrence Erlbaum.

Yosso, T. J. (2005). Whose culture has capital? A critical race theory discussion of community cultural wealth. *Race Ethnicity and Education*, *8*(1), 69–91.

Zack, M. H. (2000). Jazz improvisation and organizing: Once more from the top. *Organization Science*, *11*(2), 227–234.

6 Qualitative Sound Methods

Intentional Oscillation: Getting Attuned to Sonic Methodologies

It often feels as though I found my way into sound method as a process of intentional oscillation, moving back and forth between a lifetime in music, attending to education and scholarship with an ear to the ground, and the result of trying to apply some of the studio recording practices I experienced as a musician to conducting research. In retrospect, my work with sound practices in many ways began as a doctoral student when I was busted by a fifth grader for having a sidebar conversation during a classroom lesson during one of my first forays into qualitative research. It was the kind of moment that is not uncommon for those who study daily interactions in schools: a student says something interesting and we quietly continue the talk, trying to stay below the sonic surface of an ongoing lesson. It was this experience that led me to try and notate a minute of classroom talk like a musical score presented in the previous chapter, using two frames in Excel for every second of talk (Gershon, 2006).

Not too much later, I tried something that seemed like a good idea at the time, letting participants borrow my recording device and talk to each other without me present. Over the course of my dissertation, this practice yielded some of the most important insights from the children whose lives I considered in an effort to better understand students' academic and social roles in schools. Then, I mainly relied on a minidisc player, a device that allowed me to mark different parts of a conversation after they had been recorded like a series of ongoing tracks on a CD. Upon handing the device over to the fourth and fifth graders in the study, I would tell them something like: "This is the microphone, it's pretty sensitive so there's no need to bend into it and yell, it'll hear you even scratching your head. Here's the minidisc player, please don't drop it." I'd then sit on the side of the yard during recess watching kids record whatever came to mind: the sounds of their friends playing a game, a conversation, the squeaking of a swing.

Listening back to those recordings gave me some of the greatest pleasure of my dissertation experience, kids being themselves without me there, possibly, momentarily, forgetting that I was going to hear those recordings. It was a process that helped me better understand not only what interested this particular group of children on the cusp of being tweens but also feelings, ideas, and experiences about themselves I wouldn't have otherwise learned. To better ensure their privacy was respected as they saw fit, I worked hard to keep private information that could have hurt or embarrassed any of the kids and checked to make sure they were okay with what I used and shared publicly from those particular recordings.

It was also at this time that methodological suggestions from my adviser may well have become a more useful tool than she imagined. She told those of us who took her methodology classes, and again during smaller meetings with her advisees, that we should live our data, listen to it all the time, when you're driving, working out, whenever. While conducting my dissertation, I found that transcribing people's words, although helpful in that it allowed me to more quickly process sounds-as-text and in that I could look at exactly what someone said, lacked what I called, in my head, the soul of the matter. After listening to people talk, the sounds that surrounded and contextualized their talk, and the tone and tenor of conversations, my written transcriptions felt somehow empty.

So, instead, I began what I thought of as targeted transcriptions. By this I meant that I would listen to the recordings ad infinitum, create new tracks on the minidisc at the push of a button at what I thought to be an important event or moment, and keep track of the content at that new track mark by what is otherwise an often irritating aspect of a word processing program, the auto-numerical generation aspect of its listing function. This allowed me to add an indefinite number of tracks to any recording and keep up with them by adding titles in a way that didn't erase my previous heading and its accompanying timestamp on the recording. Then, rather than transcribing entire interviews or classroom conversations, I would transcribe only a particular section. I hadn't thought about it this way at the time but I was well on my way towards working methodologically in, with, and through the sonic. And, as I was hard at work moving through the dissertation process, there was increasing international attention to sound as a field of study from multiple perspectives and disciplines.

Sound studies began to assert itself as an emerging field in the first few years of this century, a movement that turned discussions about sound from disparate fields—ethnomusicology, architecture, anthropology, sociology, physics, art, musicology, cultural studies, etc.—into a unified grouping. This meant that what I had been doing could be classified as scholarship that at once provided me an opportunity to continue working in my home field of education and within the developing parameters for thinking about, with, in, between, and through sound.

What was often missing, however, was talk about sound in or as methodology. Beginning around 2007, I was making recordings with multiple microphones throughout a single space, then placing those sound files in a music production program so that I could listen to more than one sonic event at the same time. Doing so allowed me to hear what was happening in the background of one area of the room and the foreground of another simultaneously, hearing what was a murmur on one recording with clarity in another and vice versa. Through this process and a slight tweak, panning the sound to a certain degree left or right that caused each track to occupy an overlapping but (slightly) more distinct placement and directionality, I found I could listen to up to four recordings at a time.[1] Listening to more than one track allowed me to hear the kinds of ideas and ideals that floated through classrooms and also between classrooms—in the same way that I could listen to four recordings of a single room, I could also listen to a recording from each of the four sites for the study. And that got me thinking about the possibilities for sonic methodologies (e.g., Daza & Gershon, 2015; Gershon, 2011, Gershon and The Listening to the Sounds of Science Project, 2012).

More specifically, my work in sounded methodologies grew out of a longitudinal collaborative ethnography. Grounded in a desire to have students be actively involved in recordings, practical considerations related to conducting a multi-sited, multi-classroom longitudinal ethnographic project, and my growing discomfort with the ways that power relations often work in all forms of educational research, participating first, fifth, seventh, and eighth graders and their teachers could participate as co-researchers. To these ends they interviewed each other, made audio recordings keeping track of their processes and embedded them silently into the tracks they made (when they remembered), and occasionally video recorded their processes. Teachers helped catalogue this information, offered their insights, shared what students had been up to since my last visit including any particular events that they felt were important to share, and occasionally documented what students were doing. Students and their teachers also were helpful in offering input on things I'd missed, adding to the conversation as they saw fit from their perspectives on recordings and findings, and approving the finalized recordings for public audiences including the sound installation described below—a work that went on tour to each participating school with multiple listening sessions as well as a grand unveiling so that those with less transportation to the museum could nonetheless hear their contributions to the piece.

It is therefore perhaps no surprise that my first methodological foray in sound was something I called sonic ethnography. In short, just as visual anthropology foregrounds pictures and other visual recordings and representations, or ethnographic film is ethnography-as-film, this is ethnography-as-sound (for another argument along these lines, check Feld & Brenneis, 2004). It follows the same conventions as contemporary ethnography with attention to questions of transparency, ethics, power, and process, and uses many of the same information-centric conventions towards a sounded expression of local understandings-in-action that simultaneously explicate local and more global sociocultural norms, values, processes, and ecologies.

In an effort to push this possibility to an almost fully sonic representation, I put together a piece of sound art that consisted of four pairs of speakers through which I routed the sounds of each classroom (Gershon and The Listening to the Sounds of Science Project, 2012). These sounds were organized so that they played simultaneously, with representative sounds playing so that you could stand in the middle of the room and have the sounds of the whole project wash over you or walk closer to one pair of speakers to better hear a particular group of students/classroom. When a song participants wrote came on, the collective volume went down and the room with the song volume was increased so that the song could be better heard.

The piece was exhibited for three months at the Akron Art Museum in an area where kids, their families, and the general public could access without paying a fee. As with most exhibitions, the names of all participating children were listed as co-authors by classroom, a large explanation sign had a few sentences talking about the project and asking audience members to think about whether kids were failing the tests or tests in science were failing the kids, and the museum's first iPad used in an exhibition sat in the back with a pair of headphones so that people could see a few video clips of the project in action, hear the songs that students had made, or leave a note for all of us.

This project led to articles on sounds as educational systems (Chapter 2), sonic ethnography (e.g., 2013a, 2013b), sonic cartography (Gershon, 2013c), sounded narratives (Gershon & VanDeventer, 2013), and, three pieces of soundart (2010, 2011a, 2011b, 2012). My purpose here is to explicate some of the technical details of these processes, the kinds of ethical reflexivity required for soundwork, and why they merit separate categorization, while attending to some of the similarities and differences with what is in many ways a resurgent golden age for what might be called narrative radio.

The sonic methodologies that are the focus of this chapter and utilized throughout are qualitative. There is a good deal of information about how to use sound to measure aspects of environments and their health, the impact of sound on bodies, and how sound might be measured as aspects of physics and studio recording for engineers. Although aspects of each are present in this chapter, my focus is on how literal and metaphorical applications of sound might be useful for qualitative researchers. Similarly, while the majority of the sounds in this chapter and throughout the book are education-centric, these understandings can be applied to qualitative research across fields, disciplines, and constructs, adapted to the specifics of orientation as needed.

The remainder of this chapter addresses some of the central processes and guiding understandings that I have employed in sound methodologies to date. Instead of a set of hard and fast rules on what does or does not make a given process something that might be conceptualized as a "sound method" in qualitative research, it is my hope that outlining the kinds of ideas and ideals I have employed to these ends might at once a) provide a general set of understandings about the types of considerations central to sounded methodologies and b) serve as a springboard for future sound explorations in qualitative research.

Not Everything Is a Soundscape (Maybe)

Not everything is a soundscape. As introduced earlier in the book, a soundscape is a term coined by R. Murray Schafer and refers to a specific set of sounds that are deeply linked to acoustic ecology, a set of understandings that privileges "natural," non-mechanical sounds over those that are "man made," and is indelibly connected to scholars at the World Soundscape Project in Vancouver, British Columbia (e.g., Truax, 1978). Yet, in their excellent discussion arguing for a sounded anthropology, David W. Samuels, Louise Meintjes, Ana Maria Ochoa, and Thomas Porcello (2010) make a strong case for an expanded definition and scope for soundscapes that includes human voices and conceptualizations of time beyond recording duration. However, even this discussion suggests that, although soundscapes might have some means for speaking to a sounded anthropology, anything that might be called a sounded anthropology might indeed move beyond such expanded parameters for soundscapes. The same call for a sounded methodology had been made in an interview of Stephen Feld by David Brenneis (2004) in *American Ethnologist* half a decade before their work. In other words, making a soundscape is not necessarily sonic ethnography and, although a sonic ethnography can be a soundscape, that is not a necessary, defining factor. What, then, are the boundaries between soundscapes and sounded methodologies?

One answer is more technical in nature. If a soundscape is any naturally occurring ecology without its person-constructed aspects (e.g., Qi, Gage, Joo, Napoletano & Biswas, 2008), an undeveloped forest versus a housing complex for example, then a large number of environments are not soundscapes. However, if one expands the definition of soundscape to mean the sounds of a given ecology regardless of its construction, an understanding that includes not only those that are typical but also sounds that are rare regardless of their frequency, then the difference between a soundscape and a sound method in many ways becomes a question of intent. Where a soundscape is the sounds of a given environment, sound method is what is used when one studies, records, or considers a soundscape.

What is helpful about this particular differentiation is that it implies a central aspect of both soundscape and sonic research methods: Sounds are always mediated and in relation. From this perspective, what one hears or records as a soundscape is particular to the person or microphone and dependent on what she can hear or record. As presented in the previous chapters, the sonic is mediated not only by the most individual of factors—degree of hearing, the sounds to which one attends, how one's body is shaped, and the like—but also by ever-increasing layers of sociocultural norms and values, the ideas and ideals that shape how and what one hears. Hearing is also contingent on the physical media through which sound travels and that interrupts or redirects its movement, air, walls, water, etc. (not to mention other complicating factors such as states of awareness like sleep and other such possibilities).

This also raises another important aspect of doing sonic research. The "sound" part of sonic research tends to operate at one of two levels, literal and metaphorical, and often bounces between both over the course of a given study or publication. These two understandings tend to be then split yet again into another set of possibilities, audible or imagined. Audible sounds are those that one can hear for one's self, an understanding that is as applicable to recorded sounds as they are to those one hears without the aid of recording and playback. A central advantage to recording is that it takes a liminal experience, hearing in the moment, and moves it to a more permanent status, though certainly in degrees rather than perpetuity (for more on this, check, for example, Sterne, 2003). That more permanent status of recorded sound creates possibilities that don't exist for non-recorded sounds. It's not that recording makes the experience of hearing more permanent, for each is yet another liminal sonic experience, but that recording enables multiple iterations of listening to the same sounds. This can also be a disadvantage of the recording, that what is available to hear is less nuanced and complex than unrecorded sounds.

How one records and the kinds of things to which one attends while recording are therefore of paramount importance in doing anything that might be considered to be sound methodology in qualitative research. I'll return to this point in greater depth momentarily but wish here to express that the act of recording, what and how one records, and the equipment, time, and effort for playback are deeply determining factors in doing qualitative sound work. This is also another answer to differences between a soundscape and a sound method. Where a soundscape is placing microphones to record a particular ecology, a sound method is using that recording towards particular kinds of study. One can record a soundscape and play it for others, even as a form of soundart, in ways that are not a qualitative sound method for research. Conversely, a soundscape is also an intentional recording with attention to specific aspects of the sonic, again muffling a boundary that might otherwise seem distinct.

This also points to another significant difference between soundscape and sound method. A sound method employs a set of previously enunciated practices that should (must, in my opinion) include the transparent discussion of the ideas and ideals that inform the purpose of study, a clear question or questions that shape direction(s) of study, the steps one will take in gathering the information to study, processes of analysis through which that information will be filtered, and the reporting of those findings so that others who were not present can nonetheless follow one's process, consideration of information (regardless of its definition or wording), and its expression. In other words, although soundart, sound narrative, radio, and soundscapes can be central to sound method, the inverse is not the case. Where a soundscape, more broadly conceived, is always present, its use as research methodology requires study, theory, some degree of procedure, ethics, and the like that many forms of soundscapes do not.

What Is a Sound Method?

What are the distinguishing features of a sound method for research and how are they distinct from other possible methodologies and processes? To begin, sound methods foreground the sonic. Although this might seem self-evident, a sound method need not be expressed in sound. Sound can be employed metaphorically or through other media, textual analysis of sound as graphic composition or art for example. While my own preference tends to be in sonic expression of information—and there is a strong case to be made that the difference between sonic methodologies and all other methodologies lies in the use of sounds to express research findings—limiting sounded-ness to a more literal interpretation has as many pitfalls as an insistence of sound-as-metaphor, history, or any other interpretation. In light of the hundreds of qualitative research methods (e.g., Given, 2012) currently employed and being created, there are seemingly endless possibilities for the use of sound-as-method, limiting how sound can be expressed or the ways in which it could be the focus of research methods undermines a move to the sonic.

This, then, requires some attention to what it means for sound to become a qualitative methodology. Here too the answer is at once open and relatively straightforward. A research method is a process through which one considers a question through a sustained focus according to prescribed procedures that include some form for analysis of the information gathered and the expression of what one has learned to others, regardless of its media, mode, or audience. From this perspective, a research methodology is a particular kind of empirical study regardless of the kinds of information on which one focuses, how it is analyzed or what counts as analysis, or the means of its expression.

For a good number of contemporary qualitative researchers, the "prescribed procedures" part can be a stumbling block and therefore requires some further discussion. By this phrase I am not arguing for processes that must always function in a particular fashion or without flexibility, nor am I incidentally insisting that procedure cannot be emergent. Instead, what I am suggesting is that even when a methodology emerges from a particular study or research project, it does so in the process of following another set of procedures that somehow fall flat or open up another set of methodological possibilities.

For example, when I arrived upon what I call sonic ethnography, I did so over the course of conducting a longitudinal collaborative ethnography. The central question of this study was whether writing songs about science might help bridge race and gender gaps in science education, a question that changed in light of the information gathered to what we called "listening to the sounds of science."

The longitudinal collaborative ethnography was intentional. I was working with four classrooms in three schools where students and teachers functioned in many ways as co-researchers, making audio and video recordings of their processes, interviewing one another, and taking pictures of one another regardless of my presence. Students and teachers also had the opportunity to participate, with the necessary layers of assent and consent, as named participants so that they could receive the credit for their work. Originally planned for two years, the four participating teachers elected to make this a three-year study then informed me of their decision, one that they extended yet another year.

Finally, the work was ethnographic, adhering to contemporary understandings of the methodology that attend to questions such as those of transparency, power, and ethics: a longitudinal, collaborative ethnography. Although there are many different approaches to ethnography (Atkinson, Coffey, Delamont, Loffland, & Loffland, 2007), the point about prescribed procedure stands. That the ethnography was collaborative and longitudinal did not remove either my need to wrestle with a complicated methodological history or to consider contemporary understandings about what ethnography means and how it is to function.

It was over the course of this project that I realized I had begun to take less field notes and spend increasingly more time listening to recordings. The tendency was particularly pronounced when I had multiple microphones placed throughout a classroom on a given day, creating the context described in the introduction where I would listen to multiple files at once so that what was background in one recording was in the foreground of another and vice versa. I then always had multiple microphones around a classroom where I was working, small handheld digital recorders that could be easily placed on a desk, the microphones students used to record the songs they were working on employed as a means for them to insert a muted track about their work on a given day, a large stereo condenser microphone to record the overall sounds in the classroom.

My transition to a much more sounded method emerged in a largely organic fashion. I moved from a) taking many notes to spending my time primarily noting microphone placement, general movement in the classroom, and other larger aspects of research, to b) spending much more time either listening as the day unfolded or speaking with participants about their processes, ideas, and feelings, conversations that were also audio recorded. As an added benefit, and one that has often been the case since, audio recordings made students in this study less self-conscious than video cameras and the small stereo recorders were easier to place than cameras to record information about what was happening in each classroom. Further, the ability to play sound recordings from different contexts simultaneously was helpful in determining what aspects of a context pertained more to the local complexities of that ecology and which spoke to increasingly less local sociocultural norms and values.

However, without a governing set of procedures, not only would such emergent changes have been less noticeable but also they would have been much more difficult to employ methodologically. For example, audio recording interviews and interactions is a longstanding aspect of many qualitative research methods including ethnography. Recording interviews did not change the more open-ended ethnographic style of interview designed to give the interviewee greater say in the direction and content of interviews (e.g., Spradley, 1979; Heyl, 2007), nor did it alter my approach to ethics, respect, or care, a point I address in greater depth below. Instead, what changed was how I foregrounded audio in my *research* process, particularly in terms of analysis, neither of which would have been possible if I had not operated with a strong sense of processes from the onset of this study.

Additionally, deciding how one will conduct research helps avoid bias, missing the forest for the shiny tree one most likes or is inclined to notice. Working from a guiding set of procedures also creates a more ethical context in which a researcher operates for it provides a means to share how one will proceed and research expectations for potential participants. Knowing how one will conduct research prior to beginning a study is a central factor in transparency, or what quantitative researchers often think of as internal validity, the ability to demonstrate that you did what you say you did or that your findings are what you claim them to be. Prescribed procedures are central to methodology precisely because they are an important factor in the difference between doing research and making shit up.

Doing work in sound does not alleviate researchers from this central aspect of qualitative research methodologies. One must keep in mind the ideas, ideals, ethics, questions of power, theories, ways of beingknowingdoing, and practices that are part and parcel of research methodologies. For example, taking sounds and creating cartographies in/as/through sound neither removes one from having a strong foundational, theoretical conceptualization of the ideas, ideals, and literatures that inform one's particular project nor allows one to just record some random sounds and call it a methodology. Rather, in addition to attending to associated literatures in critical/human geographies, education, and qualitative research practices, sound methods require additional layers of attention to associated fields where sound is theorized, practiced, and expressed, those of sound studies, (ethno)musicologies, audio engineering, and the physics of sound for example. Sound theories, sound practices.

Once the information gathering stage of the process is complete, then analysis begins in earnest, checking against understanding that emerged over the course of gathering information and reexamining information from top to bottom. Sound analysis, figurative or material, called analysis or redefined and otherwise labeled, requires working beyond textual metaphors and processes. It is no longer enough to say that an object is framed or that a researchers' gaze is central. You must take into account the kinds of possibilities expressed in previous chapters and take them seriously. When mishearing is a baseline for sound communications, what does it mean that something is unintelligible? How are possibilities shaped when one idea bleeds into another and distinctions are as much about the limitations of sustain as they are about intent? What is gained and lost in deeply listening to sonic repetitions that would have been lost had they not been inadequately maintained through an artificial recording process[2]?

Finally, there is the expression of content. When will text suffice and how much sound is necessary? Conversely, is text needed at all or would the information be better conveyed solely through sound? What does conveying information sonically do for the people and ecologies the research is seeking to express? How does the use of sound speak to the research question, information gathered, its analysis, and the integrity of the process as a whole? Is literal sound necessary in a sonic methodology? Looking across three methodologies I've proposed to date, sonic ethnography, sonic cartography, and sounded narratives, the following section attends to the reasons and rationales for the methodological choices I've made and how those possibilities have changed over time.

Reasons and Rationales: Sonic Ethnography, Cartography, and Narratives

Although my understandings of sound methods emerged organically, their foundation lies in deeply held beliefs about methodologies and their functions, some of which were described in the previous section. While there are indeed a good number, at times what feels like an overwhelming number, of books about qualitative research, from the traditional to what I call the post-nexts, some further discussion about how I conceptualize qualitative research is necessary in order to better situate the methodological choices I have made in sound methods.

For example, unlike a number of qualitative researchers, I believe validity to be a canard and generalizability to be essential for qualitative methodologies. This is because the term "validity" carries with it a need for an event or phenomenon to be reproducible with extremely exacting parameters. Qualitative research is designed to get at certain kinds of processes and their recursive iterations rather than demonstrably measurable sameness. As a teacher, I might tend to speak to students in a rather similar fashion in a similar kind of moment, more directly at moments of transition between activities for example, but am not likely to speak in precisely the same way. For this and other similar reasons, measuring the intricacies of these iterations will render them even less similar in spite of their overarching commonalities. People's daily lives, experiences, and interactions are not likely to be reproducible in the same way that a chemical reaction is and changes in context deeply change how people "be" in ways that many chemical reactions do not. Where adding baking soda to vinegar is going to cause the combination to bubble and roil regardless of how pissed off one is about a clogged drain, being pissed off as a teacher deeply impacts how one delivers instructions to students and how students react.

In addition to reproducibility, generalizability, as well as internal and external validity that I will address momentarily, is a key component of quantitative research methodologies. The argument against generalizability in qualitative research tends to follow discussions about how qualitative research is about the quality of things rather than the consistent measurement of things across iterations. Where an entire narrative study could be based on the intricacies, contexts, and possibilities of three or four people's stories, quantitative research often requires a good number of iterations to ensure that findings are statistically significant both within and across multiple iterations. Yet, if one cannot see the broader picture in a narrative (for this example), then what is the point of telling it?

Two possible ways for understanding generalization in qualitative research are that a) the specifics of a study are conveyed so well that the reader thinks to herself, I know that context/I believe it to be the case or b) how those particularities relate to other studies or information. In the first instance, it is the reader who generalizes (Page, 2000) and in the second, the qualitative researcher situates their specifics into a broader narrative. In either case, both generalize. It is this generalizability that also speaks to external validity. By demonstrating that a given study speaks to broader sociocultural norms and values (for example), the researcher is attending to questions about how a particular study fits within a broader context beyond that which was studied.

Rather than internal validity—a practice that requires a measuring exactitude that is neither helpful nor necessarily part of many forms of qualitative research—I have arrived, over time, in agreement with the notion that transparency should be a central guiding factor for qualitative research practices. To be as transparent as I intend requires two key facets. First, there is the question of whether or not I was clear about my question and process from the onset of the study. This is of paramount importance because of what Michael Agar (1996), borrowing from Charles Pierce, calls abduction, how a research question changes in light of the information one gathers. As he argues, one must have a research question that serves as a guiding trajectory for research and that question must change in order to have a successful qualitative study. The above example of how the study through which I developed sonic ethnography changed from one about songwriting to one about deep listening is a case in point.

Second is the process itself. It is of utmost importance to lay out, as much as possible, not only the twists and turns of the research process as it unfolded (as opposed to how one thought it might go) but also how those inversions and machinations impacted the research process. When all is said and done, the audience should be able to have a sense that the findings are possible from how information was gathered and analyzed through to the expression of those findings, regardless of their agreement with either what or how those findings are expressed.

All this seems to be rather traditional in our contemporary post-qualitative, entangled moment. However, I would argue that these aspects of qualitative research are central regardless of their liminality, the direction of trajectory, or the impossibility of their entanglement. Just as post-modern and post-structural understandings do not necessarily mean that they are not inclusive of modern understandings in some fashion, or post-human does not include the human, post-qualitative is not necessarily beyond-qualitative. For example, a de-centering of either the Anthropocene or the human as an acting subject does not relieve one from needing to be transparent about one's question, how one gathered information, or how one considered the information gathered.

That I speak of information gathering rather than data collection is another case in point. It is a move that both reflects my agreement with calls in post-qualitative research and my own concerns about the ethics of "data." As a researcher who studies people, what makes their lives data is my presence; before I showed up they were people engaging in daily life, after my arrival, they're suddenly data. Further, because it is the researcher who determines what is and is not data, such a determination seems to be in no small degree unethical as it removes those whose lives I'm seeking to better understand from the processes of trying to come to those understandings about their lives. Talk about gathering information not only moves away from some of the lingering historical problems in social sciences that in many ways created "race" and essentialized those with less power and position (for more, check Blanchard, Bancel, Deroo, & Lemaire, 2008; Visweswaran, 2010) but also provides a language that de-centralizes the human in favor of the multiplicity of possible interpretation of "information" in or across ecologies. However, thinking of qualitative research as a process of gathering information does not remove my responsibility for being candid and direct about my information gathering processes.

Along similar lines, I now write about the expression of findings rather than their reporting. Unlike reports that are so often written, putting a premium on texted, ocular scholarship, the term "expression" is more akin to the visual and performing arts in ways that announce this connection and imply multiple avenues for conveying what one has learned through a research process to audiences. That they are findings, even if sonic, as "findings" is a means to avoid the critique from post-qualitative researchers that analysis does not lead to interpretation. This said, my own understanding is that analysis (or review or attention or however one describes the process of systematically and systemically considering the information gathered through a predetermined process no matter how organic the application of that process across aspects of research) is necessarily an interpretation, bringing one's own perspective to the consideration of information gathered in that study.

Last but by no means least, either in terms of a more complete discussion of qualitative research practices in general or in my discussion here, are interrelated questions of ethics and power. My conceptualization of power is at once informed by scholars such as Michel Foucault (1978) who famously declared nothing exists outside of power and Sherry Ortner (2006) who argued that power is not necessarily so pervasive as to deprive people or groups of their available agency irrespective of the depth of their oppression. Regardless of one's perspective, power is central to all (qualitative) research processes, and explicit attention to the many ways it operates and informs one's practice is essential to how one conducts, analyzes, and expresses findings to others.

One way to conceptualize ethics in research is as the ways in which one attends to questions of power, respect, and care throughout the research process. For me, and in keeping with the potential of the sonic, notions of ethnics often come down to questions of dignity and care. Even when expressing a strong critique of an idea, ideal, norm, value, feeling, or practice, etc., it is essential that researchers consider and report how participants conceptualize themselves, contextualizing norms and values, and the ecologies they negotiate as directly as possible. The point is not that one cannot voice one's opinions or that there should not be critique or that writing should be somehow devoid of passionate perspectives or that it is possible to remove bias from human experiences. Rather, the point is that one should absolutely seek to express another's perspectives in as clear and direct a manner as possible, especially because fully unbiased work is impossible and your work as a researcher is therefore always biased. This is the case especially if you seek to critique another's perspective. Not only does expressing someone else's perspective as gracefully and with as much nuance as possible give them the dignity they deserve as a person, the practice also tends to provide more strength to the critique. Conversely, working in this fashion is another way in which you can actively work to avoid empty, ad hominem attacks.

Then there are historical questions about how a method might function and the ways in which the move to sound retains central facets of a research methodology. Returning to a point raised above, sonic ethnography, by its very definition, must attend to the complicated history of ethnographic work as much as to the evolving subtleties of contemporary conceptualizations when the work is expressed. A sonic ethnography is an ethnography in sound. What makes a sonic ethnography different is that the majority of the information gathered was in sound rather than text, I utilized central constructs from sound studies in my analysis, and I expressed the work in sound two different ways, as a fully sonic piece of soundart exhibited at the Akron Art Museum and in a book in process that is as much sound as it is text, where text explicates sound and sound is embedded in text.

In order to better understand the degree to which sonic ethnography might function in a completely sounded fashion with minimal to no text, instead of first putting a book together, I made a sound installation that was exhibited at the Akron Art Museum (March 17–July 15, 2012). Having provided time and effort to this end, students and their teachers stated that they would prefer that I, as one teacher put it, "do the researcher thing," and put the sound installation together on my own. It is important to note here that, although this may seem like a lack of collaboration from a researcher's perspective, asking students or teachers to either spend more time after school hours or during class time was unethical—they already had been using a portion of that already minimal time to consistently work on this project.

After listening to hundreds of hours of sounds in these four classrooms—multiple recordings in the same context, visiting each classroom on a constantly rotating basis approximately every two weeks, and the recordings students and, occasionally, their teachers made in the time between visits—I had a fairly good sense of the kinds of things that happened in each classroom. Added to this was an understanding that local and less local norms, values, and processes are evident in some fashion in any recording of any ecology. As a central purpose of interpretive research is to convey an ecology to those who were not part of those everyday experiences, I sought to express those ecological aspects in ways that at once conveyed the similarities and differences within and between each site over time.

In light of these purposes, my first move was a decision to focus on the third year of the study. My rationale for that decision was that, by the beginning of our third year working as a team, the teachers and I had evolved as a group in ways that we not only knew how to better support one another but that I had more fully evolved in my sounded approach to research. Unlike the first year in which teachers and I were getting the hang of how we would work together and teachers were still futzing with how to incorporate songwriting in their classrooms, and the second year in which the project was evolving to even more deeply foreground sound, it was in our third year that we were all operating smoothly.

However, my choice that then led to the extension of the study into a fourth year was at the incoming students' requests. Students had watched the study unfold, were looking forward to participating, and argued that, if the sound installation was going to be produced during their year, they too should have an opportunity to participate in the study and be part of the installation. Their teachers and I agreed and the soundwork I created became focused on the final year and a half of the study.

As is familiar to qualitative researchers regardless of their method, the next step was the arduous process of sorting through recordings from each classroom individually, selecting the sounds that were typical of that ecology, those that reflected the kinds of choices and processes indicative of each classroom or context. Similar to textual processes, much more ended up on the cutting room floor than into the piece I was constructing. Part of the installation had also been determined by the audio companies who had sponsored the event through their donations of an interface that connected microphones and speakers to the computer and four pairs of speakers, one pair for each of the four classroom sites. When I was finished with this process, the sounds in each classroom had become two tracks, one for each speaker.

I then placed each pair in the same audio project, plugged in eight speakers placed in different areas around my chair in my small home office/studio combo space so I could better have some separation when playing multiple pairs simultaneously, and listened to each stereo pair individually and in various groupings. In keeping with the messiness that is central to transparency in contemporary qualitative research methodologies, I elected to start the piece with each teacher introducing me to their respective students at the beginning of the third academic year. The collective whole also ended on a song sent to me from the first grade classroom of the entire class singing a song together that they had written as a class—a group song for an ending to a collaborative sonic ethnography. From that point forward, I let each classroom play the way I had originally arranged each piece before bringing them all together. The only other alteration I made was to ensure that when each song appeared in a given classroom/speaker pair, I lowered the volume on all of the other tracks and raised the volume of the recordings for the speakers playing a classroom's song. The piece ran just about two and a half hours on a loop from the museum's opening to its closing every day, run behind a closed door in an adjoining room by a computer and audio interface that sat next to an external hard drive with a backup of the installation as a precaution.

The final installation had an explanation of the piece, the names of all participating students as co-creators listed by classroom along with their teachers' names, a small plaque by each pair of speakers indicating which classroom's sounds were being played by that particular pair, and a note with a warning for younger audience members not to put their ears directly in front of the speakers. An iPad with all of the songs students made over the four years of the study and a few short clips of students working on their songs along with two pairs of headphones sat at the back wall and rounded out the installation. I included the iPad so that those who had missed the students' songs or wanted to hear them again could do so and so that there were more concrete examples of students working for those who might want to see that process in action. Each of the five video clips was under two minutes long and served, like the songs on the iPad and the short text talking about the project, as a means to explicate the sounds rather than the inverse.

This iteration has since been rendered into half of another version of a sonic ethnography. What was once the sound installation has become all of the associated sounds in this work, both as sounds that stand without text and as sounds embedded within text that, in turn, addresses the facets of the study that are not addressed in sound. The texted portion in this case provides further explanation of sonically conveyed information. Each section stands on its own, so that one gets a full sense of the project from only listening to the sounded portion and the texted portion and the sounds embedded there also articulate other key aspects of the study. This functions in much the same way watching a movie is complete unto itself and scholarship about that film covers important topics in ways that further explicate the film and can stand on their own, as media or film critique for example, but are to some degree incomplete without having seen the film. Just as the sounds are manipulated to convey a particular polyphonic narrative, cut, pasted, and adjusted in volume for example, so too is the text, another fulfilment of my ethical responsibility to participants and audience alike to transparently detail my practices and findings in ways that explicate and complicate sounds. Although this may seem overly complex, it is what you have been doing so far, tacking back and forth between sound and text.

These choices are also evident in two articles that also outline sonic methodologies in qualitative research, sonic cartography (Gershon, 2013), and sounded narratives (Gershon & Van Deventer, 2013). Where the article on sonic cartographies introduces participants' talk throughout the first half of the piece and then uses them along with similar kinds of sound files to create a sonic cartography, the piece on sounded narratives outlines the context and method followed by a soundwork that performatively articulates the methodological possibilities outlined in the article. In both cases, the articles draw on disciplinary histories and practices, in critical geography and narrative as research respectively, and transparently adhere to these traditions and their use inasthroughof sound.

Before continuing further, however, it is important to address a few questions I get with a good deal of regularity when talking about sound methodologies or presenting soundworks. First is an assumption or mishearing about what it means to work in sound. This question is often framed something like, can you please talk a bit more about how sound is more truthful than text? Sound is not more truthful than text; it can do things that text cannot. Embedded in this question are assumptions about the primacy of text and often implied are somewhat contrasting beliefs about the use and role of the sonic: Sound is for recording, not a medium for the expression of serious or legitimate research. Aside from the privilege of avoiding the need to justify what makes text a good or superior medium for conveying ideas, education continues to be primarily a sonic affair. The notion that sound expressions of knowledges are somehow inferior or anti-intellectual is at least to some degree elitist and marginalizing, another limb of settler colonialisms that privilege the written word over oral traditions for example. Not that written words, regardless of language or media, somehow aren't a powerful way to convey possibilities but instead that they are not superior to other media or sensory expression.

The next question is about degrees: If sound is so helpful, then why use text? As I've covered this question from several different directions so far, instead of reiterating those arguments, I'd like to make a small addition here. Scholars already often use other media embedded in text, why is its inversion such a stretch?

A final question I often receive is from people who find the idea of working inwiththrough sound an exciting possibility and wish to just "get a few pointers" or have a "new sound method." It is the case that anyone can turn on a microphone and get a recording. It is also the case that the recording one receives can be played for others. This is using sound (recordings) in one's work, something that is by no means new nor unusual in many kinds of research across the spectrum of qualitative and quantitative research methodologies.

With this in mind, another definitional aspect of what can make a method a sound or sonic methodology is that it attends to scholarly conversations about the sonic. Just as one wouldn't claim to be an anthropologist without seeking some modicum of understanding in various aspects of scholarly literature in the field, claiming to work in a sonic methodology requires attention to both a longstanding history of scholarship about sound and contemporary scholarship about the sonic. Although one's scholarly attention should certainly foreground sonic scholarship in one's own field, attention to sounds in other fields and disciplines in general and to a rapidly expanding literature in sound studies is warranted.

For example, in every one of the conferences I have attended over the past year, I have been approached by someone who wants to think with me a bit about a new sonic methodology they have developed. To be expressly, abundantly clear, I am very much in favor of projects like these and they are close to my heart. However, more often than not, what I hear is a slight methodological tweak, moving versus stationary audio recording processes for a not uncommon example, where the scholar in question has looked to usually ocular methodologies and concluded that it doesn't exist. Sure, part of this is a more general concern about surface or unintentionally cursory explorations of scholarly literatures. But there is another part of questions like these that is more troubling, an assumption that our relationship with the sonic is somehow either new or overly novel. This perspective overlooks radio, music, (ethno)musicologies, composing, sound art, histories of composition, dada, sounds of the spheres, soundscapes, and on and on. To my mind, this is perhaps the most clear expression of the primacy of the ocular in research methodologies. In an era where the differences between personal and professional media production is increasingly blurred, we have again arrived at an understanding that making videos is an art and that there is a marked difference between those who spend a good deal of time working with video, but audio, just throw a microphone up.

The line between professional and amateur in media production has increasingly become more a question of literal and metaphorical time and money. When it comes to audio production in music, for example, most would likely seek as professional a recording studio as possible to make a musical recording for either promotion or sale, regardless of its purpose—those who spent the time, effort, and money would likely turn to themselves.

Yet in the research world, this is often not the case. Unlike computers and phones that have rather high end cameras and associated software, the microphones associated with those devices, while certainly function-able, pale in comparison. Along similar lines, although "USB enabled" microphones (those that can plug directly into a microphone or mobile device) are often heads and shoulders above internal microphones on most non-audio-oriented media devices (those not designed for audio recording), and USB microphones have improved greatly, a noticeable difference between USB and non-USB microphones often remains.

Practically, this means that audio recording can be expensive. Even if one has a computer or mobile device and an application that is capable of recording, audio recording beyond using the internal microphones on computers, phones, and tablets often requires an audio interface, what translates the sounds/signal to digital and back again. That is if one is not going directly into a mobile field recording device, or a more analog process with tape, not to mention the wide variety of kinds of microphones from which to choose and their respective specifics and price range. And doing sonic work tends to require getting good sound; and getting good sound requires good equipment and an understanding of how to use it.

This then is the last central point of this section: Doing work in sound requires more time, effort, cost, and, for lack of a better catch all phrase, skills, not less. The hundreds of hours listening, analyzing, and manipulating[3] sound that I spoke of above are not hyperbole. In order to fully engage in sonic ethnographic work meant that I needed to be similarly additive in my methodology as I was in my literature review—my process was the painstaking, inter-convoluted work of ethnography compounded and intersected by multiple traditions of doing work in sound and sonic scholarship. What remains is talk about the more technical aspects of how I used sound in these three sounded methodologies and my reasoning for making those practical, technical decisions.

Getting Technical: Sound Recording and Production

That I was and remain drawn to sound is not surprising given my lifelong love affair with music and all things sonic. As a working musician, albeit with much less frequency in the slowing combination of working to make tenure while living in a small Ohio town, I have spent a good deal of time around recording equipment in both live and studio recording contexts. Similar to many musicians, I also have a small home recording studio and a few microphones geared towards recording the instruments I most often play, saxophones and Afro-Caribbean percussion along with associated smaller instruments like shakers that are often collectively referred to as toys.[4]

This combination created a context in which I had some familiarity with recording processes including the use of music production software and audio engineering. However, it is important to note that what might seem like a false humility about my experiences is in fact actually the case. While I did indeed have a fairly good sense of such aspects in audio recording processes as how recording programs function, some understanding of microphone placement, and the ways in which various recording effects (e.g., reverb, delay, and compression) can be employed, I did not study or apprentice as an engineer and this is not how I make my living at this time. There is a reason that someone employs full-time audio engineers and studios, regardless of their level of fame, and noting those differences is of no small significance in an era when the tools of the trade are now so readily available and employed.

Becoming Audible

One way to consider the audibility of a recording is in two overarching stages, recording and production. In keeping with central tenets of qualitative research such as transparency and interpretation, I have consistently made choices that cut against expectations in music production. This section details the rationale behind these and other such choices and describes some general practices that others might use for their own recording. When thinking about sound, it is of utmost significance to note that, far too often, researchers spend a good deal of time attending to the visual aspects of a video recording and far too little time on the audio. It is a difficult lesson I have watched emerging and even experienced researchers rediscover the hard way: excited to watch a particularly meaningful interview, interaction, or ecology only to find that it's beautifully framed and almost entirely inaudible. Although this is indeed a byproduct of the primacy of the ocular, the point I hope to make here is more practical in nature, the discussion of recording choices in this section applies as much to video recording as it does to audio (only) recordings.

Microphones matter. There's only so much a person can do to make a recording sound better or make the inaudible clear. Regardless of the kind of microphone one selects, microphone placement is essential. One cannot just put up a microphone or group of microphones and assume everything will be okay. This may seem a bit pedantic but I've been in too many contexts when it didn't happen. Before recording, test the microphone in a variety of settings. Find what is too close or too far and how a microphone handles sudden, loud sounds and whispering, far away sounds. Take care here too as microphones can be damaged by sounds as easily as they can do their job vibrating to record the sonic. Always have an extra microphone in all recording contexts. The number of times I would have lost all information from an interview or an ecology without a suddenly-not-so-redundant microphone is not small.

Many small handheld digital recorders are now available with stereo microphones. The upside to using these recorders is that they record in stereo which is, of course, the downside to the recorders. By this I mean that these kinds of stereo recorders provide some separation between sounds and a more everyday sonic spread that's closer to how one would hear: two ears, two microphones. The downside is that these two microphones are recorded on a single track. Compare this to a pair of mono microphones placed so that they provide a stereo recording. Because the mono (single) microphones each record separate mono tracks that create a stereo recording when placed together, each individual mono track can be manipulated individually. As one can imagine, this is incredibly helpful when there is a loud sound to the left or if most of the action happened more on the right, enabling the researcher to raise or lower the volume accordingly. On a stereo recording, one can pan the sound more to the left or right but changing the volume alters it on the full recording. Conversely, matching adjustments on separate mono microphones is more time consuming. However, whenever possible, as much ability as possible to treat each side of a stereo recording on its own needs and merits is most helpful. This is also the difference between a professional grade stereo microphone and many handheld stereo recording devices. Professional stereo microphones (field or otherwise) record to two separate tracks and are, in fact, paired mono microphones. This said, a good mono microphone often does the trick.

Language used to describe various microphone tones and tendencies can sound more like selecting a cigar or a fine wine (or, more recently, a micro-brewed beer) than a piece of equipment. However, just like descriptions of wine, cigars, and beer, such distinctions matter and are as much about the kind of mic needed for a job as it is about personal preferences or appearances—a high end mic or few can make it seem like you know what you're doing and running a high end shop. The caveat? You need to deliver the goods. What counts as a "good microphone" is, as is the case for most qualifying descriptors, about specs and context, what microphone for which job, and personal preferences. There are now a ton of free, online resources about microphones and recording in general. With this in mind, here are some of the choices I've made for recording in classrooms and my reasons for working in that particular fashion.

The first decision was between a field recorder and a computer-based digital audio workstation (DAW). I elected to go with a DAW that had the most choices for the least amount of money, a choice that also aligned students' work on their computers with the program I'd be using. This decision created a context in which students could also manipulate sounds on my machine, although that happened with relative infrequency though was offered at least once a month, and I could more readily insert their recordings into files for analysis and creating soundworks. Choices I made about those recordings are addressed in the following section.

Once I made the decision to use a DAW through my computer, I then needed to get an Analog→Digital converter, the device that converts the "real world" sounds that cross a microphone element to digital 0s and 1s. There are many relatively inexpensive AD/DA converters and finding one that works is as much about cost as it is about specs. In my case, I went with a 2 in/out converter which means that I could handle two inputs at any time (two microphones or two guitar cables or a microphone and a guitar cable, for example) and the device had two output lines that could go to studio monitors, the speakers used for playback in audio recording and engineering. I also did so because I rather quickly realized that I was building a home recording studio and the device I used for field recordings could also double as the interface for playback when back at home.

For making more high end professional recordings of classrooms, I alternated between a pair of what are called pencil microphones, smaller mono microphones that are placed in various configurations to get stereo recordings, and a stereo field recording microphone that had two different mid-sized condenser microphones in x/y configuration on a single microphone stem. This means that off of the metal part of the microphone, the part that a singer holds but more slim than you're likely imagining, there were two outward facing microphones angled so that they make a triangle shape with a metal support on the bottom and each of the other faces has a microphone grill. Unlike many of the smaller handheld recorders, stereo mics of this design function as a literal two-in-one microphone with each of the pair recording independently at the same time—my stereo microphone occupies both of my two converter inputs. Over time, I tended to lean more on the stereo field microphone than the pencil pair primarily because there was little difference in the kinds of sounds I gathered and the single field microphone required much less setup time, which is, of course, one of the reasons field recording microphones are constructed.

Microphone placement matters. Even though I tested levels before I began recording and had placed the four handheld recorders I used in various situations to see how they handled sounds from different distances and volumes and impact (sudden, loud sounds for example), it still took some time to get the hang of what was going to work with them. This was because, unlike the microphones I was used to using either on stage or in the studio, or even the field recording microphone, getting a read on the volume indicator on the side as compared to the levels that would be recorded was not as consistent and there was no room to create limiters or passes that would cut sounds above and below particular thresholds. A student suddenly yelling in excitement would always peak the recording, often distorting it, and a student suddenly whispering might end up being all but unusable in spite of the ability to raise the volume in the studio.

However, once I had a better sense of how the microphones functioned, their small, slim profile provided the means to more readily place them near students without too much interruption. Getting a better sense of how the mics behaved also made it easier to speak with students about microphone placement when they made recordings on their own. The handheld audio recording devices also tended to hold power for a comparatively long time, didn't lose recorded material when the batteries did crap out, and were rather rugged; not one broke in the four years of the study that included first, fifth, seventh, and eighth graders.

In general, it is best to place microphones so that they are close enough to a sound that they "hear" it clearly and far enough away that there is no feedback and, conversely, that sounds are not too soft to be audible. Although this may sound like the most elemental aspect of audio recording, it is also something that is often overlooked—it is not unusual for a colleague who has made a video or audio recording to find me to see if I have any suggestions for cleaning up the sound. Holding a microphone directly in front of your mouth can make all the plosive sounds of talk (the initial "p" and the "s" of plosive for example) peak the audio and holding it too far to the side means you might be less audible depending on the directionality (focused) of your microphone (holding a mic angled a bit to your right and turning your head somewhat to your left can cause a much less audible recording; ditto for recording conversations and the like).

It is next to impossible to get a non-studio recording without sounds other than the ones you had in mind to record in the first place. As I describe below, an important difference between sound methodology and journalism is intent and purpose. In general, sound journalism has specific sounds in mind constructed in particular fashions that speak as much to sociocultural conventions of what radio programming (for example) should sound like and how it should function. Methodologically parallel to how questions change in light of the information gathered in qualitative research, sound methods work from a position that recorded sounds in a particular research ecology will provide sonic information that is significant in both anticipated and unanticipated ways and that such information will impact one's analysis as affectively as cognitively and speak as much to notions of sensation as they do to constructions of signification.

Personally, I'm on the John Cage (2013) side of things, by which I mean that silence is an impossibility and that "incidental" or unanticipated sonic information is as valuable and significant as those one intends to make. The side conversation at the next table over, barely audible at first, that speaks to a key feeling students are having in a classroom; the electronic hum of lights; the shuffling of chairs that increases as direct instruction goes just 20 minutes too long; street sounds as youth teach each other hanging down the block. All sounds, regardless of your intent to gather them or their ability to be categorized as talk or music, are informational and hold potential keys for better understanding that which one intended to learn. The sonic also holds an infinite number of feelings and knowledge that, while not gathered intentionally, enrich both contextualizing information and the intended question in surprising ways that can ultimately be of greater help that those sounds one intended to record in the first place.

Finally, it is always a good idea to take a quick note in prose or illustration about where microphones were placed on a given day, the timing of the placement, especially if they are moved, and where people were situated in relation to the microphone. Directionality is also important and often overlooked as where the mic is pointed is a strong determinant in the sounds that are recorded.

Ethics of Audio Recording as Research

Ethics are central to sound methodologies as they are to all forms of research. This said, working in, with, and through sound does not require more or less ethics than text or video. What makes ethical questions about sound production and expression seem as though it should require more of an awareness to questions about power and expression, for example, is the privilege and false commonality of text. Text can hide, obfuscate, and essentialize individuals, groups, and ecologies as well as any media available and continues to be used in this fashion. It is also the case that text creates a kind of affective distance that can make critical attention more readily employed as opposed to the more affectively triggering aspects of audio and video. The kinds of ethical decisions I address in this section therefore attend to some of the ways that sound can intentionally and unintentionally move away from more transparent expression. They are also applicable as much to sonic methodologies as they are to audio recordings in qualitative research more generally.

One of the first questions in working with sound is the degree to which any individual file is manipulated after it has been recorded. This can occur in the analysis or expression of findings. For example, one might take a section of a recording rather than an entire recording in the same way that one only takes one or two lines of quoted material in text. One might also raise or lower the volume in sections or altogether, use an equalizer to sculpt the sounds so that particular qualities or sound ranges are amplified or recede into the background, blend sounds so that two different selected pieces of an audio file create a generally seamless single file, or add a touch of reverb to make the recording have more sonic space than the actual physical space allowed.

Throughout all the recordings I've made either as sound installations in galleries or other public spaces and those that accompany scholarly publications, or are the publications, I have adhered to the same set of principles. Other than moving audio files, a choice that includes cutting and pasting sections of sounds, or changing the volume, I have intentionally not altered any aspect of the sonic information gathered. Cuts between files are left rough so that there is a bit of a skip or gap between each recording. The sounds embedded throughout a piece are contained in a final soundwork and vice versa, a move that places the information addressed in the hands of the audience so that they can not only hear the information for themselves but also make a more informed decision about the arguments I have made.

Along these lines, I usually do not add reverb to make a room larger but, when a sound is rather inaudible due to a high or low end hum, talk during an interview for example, I do on a rare occasion use an equalizer to lower sounds in the interfering range and boost those in the range of talk—lowering the high end of the spectrum and raising the midrange in the case of a high end hum for example. Manipulating sounds in this fashion does indeed alter sounds in subtle and more obvious ways. As a result, a central part of mellowing sounds that mask a context is to find a balance where the recorded sounds retain their particular sonorities while the less desired sounds are lost. In keeping with the arguments made throughout this book, it is important to note that these manipulations are small, subtle, and infrequent. This is the case in part because of microphone placements that, while they can and do result in sudden loud or soft sounds that require a limiting of volume but not a section of the sonic spectrum. It is also because I am most often inclined to generally leave a file alone, either letting a hum stand or electing not to use the recording in much the same way that well-crafted paragraphs nonetheless don't make it into a manuscript.

Although this may sound like a good deal of manipulation, it truly is not. Of the hundreds of available tools that can manipulate sounds in thousands of ways, I change them in exactly three ways: cutting and pasting to tell a particular story, adjusting the volume so that sounds are more audible, and occasionally cutting the high or low end while boosting the range of sounds most prevalent in a recording. That is much less work than what goes into any news radio broadcast.

Another question central to discussions of ethics in sound focuses on soundtracks, sonic, often musical, transitions between audio recordings, voiceovers, and other practices that often accompany the construction of sounded narratives. In general, I do not use sounded transitions between sections, a song as an interlude between points or audio moments for example. On the rare occasions that I do, it is something that students have created that is part of the overarching arch of a narrative, the sounds of students working in the creation of a song that culminates in the song that they wrote, for example. While voiceovers are indeed a strong tool in audio and visual expression, they often carry an extra authority that tends to supersede other expressions. For the work that I tend to do, ethically, I have already changed the arrangement and amount of audio in order to express particular narratives that are, in turn, scholarly arguments grounded in the analysis of information gathered over the course of a study. With this in mind, if I have done my job well, a voiceover is unnecessary and any additional information can be delivered via text, a conscious and explicit distancing that is more difficult to conflate with a given narrative.

Further, the addition of sounds, music or otherwise, to audio files is an additional layer of affective manipulation that further presses a point: melancholy-seeming music to literally underscore a sad or difficult feeling. This can also create a false relationship between expressions, ideas, and/or, ideals. Such a move can add irony or criticality to a particular recording and might be applied in creative ways, opportunities that can add depth and character to a soundwork. This said, I find that the concerns often outweigh the possibilities and tend not to use contextualizing sounds in this fashion. Nonetheless, I have un-muted students' recorded explanations of the instrumental songs they created so that they become a voiceover that expresses student choice in music creation, played a students' created bass line against his own talk about playing bass, and other such decisions where the pairing of sounds functioned to contextualize one another, decisions that were also checked with those involved before being shared with a broader public.

I do, however, often set two soundfiles to play simultaneously, one playing primarily out of each speaker so that listeners hear both sounds at once. This choice provides an opportunity to hear an event much more like one might in an enclosed space like a classroom, where the sounds to one side are the background noises in the other and vice versa. People have an amazing ability to discern sounds in a complex sonic environment, what one person is saying in the middle of a crowded lobby or background singers in a multi-voiced and instrumental song for example. Yet, it can take some people time to orient to this kind of arrangement in no small part because of our listening expectations rather than our daily experiences and our textual, ocular inclinations, a question of preference rather than possibility.

The point here is not that these are some unassailable ethical boundaries or that I have somehow figured out the key aspects of how sound methodologies should or must function. Rather, these are examples of the kinds of ethical decisions that are endemic to working in, with, or through sound. They are choices that are in many ways similar to the kinds of ethical arguments one makes regardless of media, possibilities that are further compounded by a sonic component and therefore require additional awareness and care for similar attention in relation to aural media. Sounds can mask as much as they reveal, the act of recording does not make something more truthful, and what is heard remains a constructed mishearing. It is precisely for these and other such reasons that researchers must establish and adhere to particular ethics that address general practices of care, respect, dignity, and awareness in research, coupled with further considerations in relation to media. For, if sounds can express things in ways that text cannot and are qualitatively different than video, then they can obfuscate and subvert things in a particularly sonic manner as well. What sound can do is to provide a means for research participants to express themselves in their own voices without showing their faces and provide the ability to hear ecologies without necessarily giving away significant identifying factors. This also attends to a central conceit in many forms of qualitative research, scholars telling, and getting credit for, other people's stories. It is without a doubt of paramount importance to protect participants from harm as much as such things are possible. Yet, the combination of protecting participants' identities while telling their stories creates a condition in which scholars, rather than participants, are venerated.

Finally, it is important to remember that sound is at once media, mediating, and mediated. It is material, a material that impacts other things and ecologies, that is, in turn, impacted by ecologies and things including that which it impacts. Then there are microphone placements, quality of gear, and many other ways in which sound is manipulated in the research process. All of which is to say that although sonic methodologies function in different and often more affective ways than text, they are not necessarily more truthful. On the other hand, sound methodologies can indeed do things that text cannot, analog experiences, often through digital playback, that speak to everyday experiences with an immediacy and complexity that can feel more vibrant.

Intention, Attention, and Expression: Sound, Art, and Radio

Interactional processes such as research and performing arts can be considered in terms of what they *intend* to do, the kinds of things that are within the scope of participants' *attention*, and how those intentions and attentions are *expressed* to others. Intentions are complex and beyond the individual and human, attentions are as potentially diverse and diffuse, and their expression is always necessarily implicit, explicit, embodied, affective, sensory, cognitive, and creative. From this perspective, sound methodologies are sonic in their intention, attention, and/or expression.

Central factors that separate sound methodologies from sound arts are the ethical and procedural obligations and congruencies present in research that are not necessarily as apparent in visual and performing arts. In saying this I do not mean that visual and performing arts are somehow devoid of obligations, ethics, procedures, or forms of congruence as this would be a ludicrous argument and flies in the face of my belief in the power and potential for the arts that I continue to voice in theory and practice.

What I seek to articulate is the following often subtle but nonetheless significant difference between arts and research. For example, in my role as a musician, it is my intent to play something that is melancholy and foreboding. My intent is the result of attention to thoughts about concerns I have about the rise of fascist attitudes that are made possible by what is an increasingly obvious lack of concern about such politics on one hand and an acceptance of them as normal on the other. I then express those intentions and attentions as long, wavering, languid phrases with "unexpected" and often "atypical": modulations in pitch, key, and tone.[5] Whether an audience member perceives my intent, attention, and expression according to the parameters in my person or not does not (necessarily) impact an audience member's appreciation, enjoyment, rejection, or dislike of that performance. There need not be participant and audience agreement or congruence for both performance and its perception to be experienced, regardless of its valuation, as positive or negative for example.

Sonic methodologies, however, like all qualitative research methodologies, do require both transparency of process and an explicit attention to an ethics. Transparency in process provides a means for audiences to consider the degree to which a research process functioned as the research claimed, a central factor in audience members' ability to find all aspects of a study trustworthy, from information gathering through to its expression. Ethically, there is a responsibility for conveying the information gathered in ways that render it sensible for those who were not present. This is particularly the case in research of the kind that I most often do, working with traditionally marginalized youth populations—the greater the oppression or marginalization of the participants, the greater the responsibility to deliver information in ways that convey participants' understandings so that they are not unintentionally or intentionally altered or obfuscated. There is a similar responsibility in analysis and expression, making explicit that which is participants' unfettered expression and that which is your alteration of that expression as a result of analysis; working in sound does not alter one's responsibilities as a researcher.

The expression of sound methods as versions of soundart, however, can provide significant benefits for researchers. This is because sound installations can function as public art and in gallery spaces, both of which grant access to the public in ways that can more deeply impact others and move outside the academy. For example, where public art can turn passersby into audiences that may elect to pause and further participate with the artwork, gallery spaces tend to provide more prolonged access to a much more limited public, those who can afford or wish to attend gallery spaces. Both have advantages and disadvantages in relation to questions of race, class, gender, and the like: Soundart in a gallery space can grant more sociocultural weight and status to a work or participants while limiting access to the work for example. Such choices are strongly informed by ethics and intention. My decision to place a sound installation in a museum of modern art was in no small part due to a desire for a more elite public to hear students of color as thoughtful, critically creative people in the face of test scores that often labeled them as failures. Conversely, the decision to place much of the soundwork associated with scholarship I've produced online is driven by a desire to reach as many people as possible, including those who do not have access to the written portions of these works. However, regardless of the location of its presentation—online, in person, in a museum, or at a street corner—soundart can do something that is often difficult for researchers, reach an audience outside of the academy.

Sound Method vs. Narrative Radio

Sound methodologies and those that Jessica Abel (2015) calls "the new masters of radio"—a group producing radio that I have come to think of as narrative radio, programs that are story-driven rather than news programs, game shows, religious programming, sports radio, or call-in programs regardless of purpose or orientation–are similar yet distinct processes. Specifically, I am thinking about shows such as This American Life, the Moth Radio Hour, and Radio Lab, programming that is often found on National Public

Radio. There are number of excellent works that consider how to tell stories through radio and about radio programs are made (e.g., Abel, 1999, 2015; Biewen & Dilworth, 2010; Kern, 2008). Although there are many different possibilities to these ends, they tend to fall along a spectrum from going out to get a story to getting information and seeing the story that emerges. However, most narrative radio programs, those that tell stories as opposed to those that are news programs or game oriented programs, are a combination of journalistic research and storytelling technique that are as much about planning and structure as they are about information gathering.

Narrative radio and sound methods have a good number of things in common. Both follow a question or theme until it becomes something different than what was originally expected or as Jay Allison tells Jessica Abel in *On the Wire*, "The phrase we use is 'the moment you know you have a story is the moment you realize it's not the story you thought it was'" (p. 66). This sounds rather familiar to advice given to interpretive researchers by scholars such as Harry Wolcott (1990) and Michael Agar (1996), that strong ethnographic work begins with a question that morphs in response to the information gathered along the way. The change can happen at any time, during the information gathering stage or during analysis for example, but is a defining facet of many forms of qualitative research.

Narrative radio and qualitative research also share a sense of planning, adherence to particular structures, a focus on making the strange familiar and the familiar strange (Spindler & Spindler, 1982), and an ethical fidelity to respecting those whose stores are shared and conveying information in a way that situates people's experiences in ways that resonate with a wide listening audience. Yet, most radio programs are more about prescribed directionality than discovery, and this prescription makes the difference between many sound methods and radio.

In many ways, these are the differences between journalism and qualitative scholarship. Recognizing that this is an increasingly blurring line, where journalism seeks out a story and moves to tell it, qualitative research poses questions and gathers information to better answer a question. As Abel conveys, this is most often done by getting information that conveys particular perspectives rather than finding what perspectives are about a particular perspective. What ends up on the cutting room floor in most forms of narrative radio are bits and pieces that move against the core story. Sonic methodologies, on the other hand, I argue, seek to find plausible means for conceptualizing the interrelations between ecologies, groups, and individuals in ways that render their movements and decisions sensible—a move that encompasses rather than dismisses what might seem on first blush to be disconfirming information. Sonic methodologies also need not reconcile oppositional perspectives to convey a narrative for narratives in qualitative research are often made stronger by the presence of irreconcilable information.

Although there are other important differences and similarities between narrative radio and sonic research methods, this central point remains: Sound methods gather information to better understand questions, and narrative radio tends to gather information in support of particular suppositions. This said, there are narrative radio programs where this tendency is not the case and there is much to be learned about the potential and pitfalls of sound methodologies from radio programming in all its forms and in all stages of production. In sum, the differences between sound art, narrative radio, and sound methodologies lie not only in their intent, attention, and expression but also in their purpose. Just as ethnography is different from collaborative action research and qualitative research is different from newspaper reporting, working in sound does not remove these or similar distinctions.

Sounding Methods

What it means to do research in sound is still evolving, and the claims made throughout this chapter are intended to serve as examples of the kinds of considerations central to doing qualitative research in/of/through/about/with sound. To begin, it is important to note that—provided that one attends to questions of process, ethics, and the like that are essential to all forms of qualitative research—there are many potential pathways for sonic research methodologies. For example, this chapter generally did not address what it might mean for music to be a form of sonic qualitative research and I limited the conversation to sound in qualitative research neglecting a great number of possibilities for the sonic in quantitative research (for example, check Qi, 2007).

Recording everything in a context and then putting it together in a soundwork is not sonic ethnography unless it follows the tenets of contemporary ethnography. Although this may seem like an oversimplification, in light of the kinds of questions I am often asked about doing sonic research, it has come to my attention that this is an important point. Just as throwing microphones up is not necessarily research, compiling it into narratives is not ethnography per se. My concern is less that there is a particular kind of research than it is that there is a research practice at all, a process that exists prior to research so that those who are not present can at once follow your moves as a researcher and you can be held responsible for your actions.

The inverse is also the case. Talking about listening, voice, silence, or hearing is not necessarily doing sonic research or methods. There is now a great deal of scholarship focusing on the sonic from a multitude of disciplines, fields, and processes. The comparative newness of a scholarly area called sound or sonic studies is no excuse for a lack of attention to associated theories, methods, and practices. Further, doing sonic scholarship necessitates sound methodologies and working with sounds requires not only benefitting from the additional possibilities of the sonic but also attending to how those complications inform and impact research processes.

Doing research that involves sound has a history that reaches well past a 20th century moment where audio recordings were not only possible but plausible. Although non-textual research has now become more mainstream, there is still a great deal of pushback against sound scholarship, both implicit and explicit. Implicitly, a good number of national organizations and scholarly publications cannot, and will not, accept any form of audio file in or as submissions. Many of these same organizations now specifically encourage other visual forms of expression, types of video and image formats for example. Explicitly, there remains a strong bias against non-textual work as "real" scholarship. Here are two examples, one local and one national. Locally, my inability to count soundart in museums and public spaces towards my tenure review when those very works would have been positively adjudicated towards tenure for colleagues in the arts speaks to the narrowness in which scholarship in many fields is understood and the primacy of the ocular in U.S. education. At a national level, discussion on National Public Radio about whether or not listening to audio books can count as reading, a point made seriously silly by calls from blind individuals talking about their experiences with audio recordings as readings, operated from a similar perspective about the significance of text and its place as the central, if not only, conduit for serious scholarship.

It is therefore in sonic expression that sound methodologies are perhaps at their most radical. Expressing research through sound turns passive readers into active listeners, requires a kind of slowness necessary to let ecologies and those that populate them unfold in real time, and an opportunity for all populations in general and traditionally marginalized populations in specific a chance to speak for themselves. Allowing people, places, and things to sound off on their own terms at once makes the collaborative aspects of qualitative research more explicit, encourages listening to others and a kind of slowness of attention, and presses back at understandings that conflate scholarship and text.

Finally, it is again important to note that affective sense-making through the sonic operates in a qualitatively different manner than a textual analysis. Like the inability to blink one's ears, the hours spent finding, adjusting, readjusting, and re-listening to sounds, all of which are also iterations of hearing, is its own form of deep listening. Working in sound not only conveys information to others but it also shapes your self with reverberations and resonances that reattune your sensibilities and recalibrate understandings as they echo long after a project is done.

Notes

1 I can pan them so that each is more prevalent in one ear than the other, more presence in the left than right ear and the inverse, a separation that mimics the binaural placement of ears and how we locate sounds through them that has been used in studio engineering and mastering for nearly 100 years. This is also the reason that jazz and pop records from the mid-1950s through the mid-1980s have the bass and drums in one ear and the guitar and keys in another, to echo how bands set up and are experienced by live audiences.

2 All recording processes are not only mediated but are also to some degree false, or, perhaps, creating a different kind of normalcy, in which otherwise liminal experiences become more permanent through their recording, a process through which as much is often lost as is gained.
3 The term "manipulating" here is also meant literally. Talk about how I manipulate sounds, change them according to sets of transparent parameters, is the topic of the following section. In this instance, manipulation is not necessarily negatively connoted as it refers to a more literal changing of sounds. This said, and as I speak to momentarily, that one seeks to alter sounds based on ethical parameters by no means absolves a researcher from attending to the multiple ways in which such changes can be harmful, regardless of their intent or the researcher's awareness. In other words, attention to such ethical questions is part and parcel of sound methods.
4 In spite of this moniker, instruments like caxixi and even those with the most simple-seeming, straightforward design like shakers are indeed musical instruments that sound and move in noticeably different ways when played by percussionists.
5 Unexpected and atypical are in quotations here as they are not necessarily either unexpected or atypical depending on one's listening habits and expectations. They are, however, intended to move in less typical rhythmic, tonal, and harmonic patterns for many forms of music, as most if not all musical forms have underlying patterns, expectations, and sociocultural expectations. Even freely improvised music has variants and subfields with differing expectations including that few pre-established patterns emerge while playing.

References

Abel, J. (2015). *Out on the wire: The storytelling secrets of the new masters of radio*. New York: Broadway Books.

Abel, J., & Glass, I. (1999). *Radio: An illustrated guide*. Chicago, IL: WBEZ Alliance Inc.

Agar, M. H. (1996). *The professional stranger: An informal introduction to ethnography* (2nd ed.). San Diego, CA: Academic Press.

Atkinson, P., Coffey, A., Delamont, S., Loffland, J., & Loffland, L. (Eds.). (2007). *Handbook of ethnography*. Thousand Oaks, CA: Sage Press.

Biewen, J., & Dillworth, A. (Eds.). (2010). *Reality radio: Telling true stories in sound*. Chapel Hill, NC: University of North Carolina Press.

Blanchard, P., Bancel, N., Boëtsch, G., Deroo, E., Lemaire, S., Forsdick, C., & Bridgeman, T. (Eds.). (2008). *Human zoos: Science and spectacle in the age of empire*. Liverpool, UK: Liverpool University Press.

Cage, J. (2013). *Silence: Lectures and writings, 50th anniversary edition*. Middletown, CT: Wesleyan University Press.

Daza, S., & Gershon, W. S. (2015). Beyond ocular horizons: Sound, silence and sonification as inquiry. [Special Issue: 20th Anniversary]. *Qualitative Inquiry, 21*(7), 639–644.

Feld, S., & Brenneis, D. (2004). Doing ethnography in sound. *American Ethnologist, 31*(4), 461–474.

Foucault, M. (1978). *The history of sexuality, volume 1: An introduction*. New York: Vintage.

Gershon, W. S. (2006). Collective improvisation: A theoretical lens for classroom observation. *Journal of Curriculum and Pedagogy, 3*(1), 104–136.

Gershon, W. S. (2009). *The collaborative turn: Working together in qualitative research*. Rotterdam, NL: Sense Publishing.

Gershon, W. S. (2010, December 17). *Sound identity: Race, space and place in an urban middle school*. [Sound Installation]. Kent, OH: Kent State University Gallery.

Gershon, W. S. (2011a, April 2). *Coming out to others, coming out to support*. [Sound Installation]. Kent State LGBTQ Minor Fundraiser for the Emergency Fund.

Gershon, W. S. (2011b, April 9). Innovative approaches to scholarship II: Listening to the sounds of science: Knowledge and making sense in urban middle grades classrooms. Sound Installation/Data Representation. *American Educational Research Association*, New Orleans, LA.

Gershon, W. S., and The Listening to the Sounds of Science Project. (2012, March 17–July 15). *Making sense of science: The sounds of teaching and studenting in four urban classrooms*. [Art/Sound Installation]. Akron Art Museum.

Gershon, W. S. (2013a). Vibrational affect: Sound theory and practice in qualitative research. *Cultural Studies ←→ Critical Methodologies, 13*(4), 257–262.

Gershon, W. S. (2013b, May). Resounding science: A sonic ethnography of an urban fifth grade classroom. *Journal of Sonic Studies, 4*(1). Retrieved from http://journal.sonicstudies.org/vol04/nr01/a08

Gershon, W. S. (2013c). Sonic cartography: Mapping space, place, race and identity in an urban middle school. *Taboo: The Journal of Culture and Education, 13*(1), 21–45.

Given, L. M. (2012). *The SAGE encyclopedia of qualitative research methods*. Thousand Oaks, CA: Sage Press.

Heyl, B. S. (2007). Ethnographic interviewing. In P. Atkinson, A. Coffey, S. Delamont, J. Lofland, and L. Lofland, (Eds.), *Handbook of ethnography* (pp. 369–383). Thousand Oaks, CA: Sage Press.

Kern, J. (2008). *Sound reporting: The NPR guide to audio journalism and production*. Chicago, IL: University of Chicago Press.

Ortner, S. (2006). *Anthropology and social theory*. Durham: Duke University Press.

Page, R. N. (2000). The turn inward in qualitative research. *Harvard Educational Review, 70*(1), 100–108.

Qi, J., Gage, S. H., Joo, W., Napoletano, B., & Biswas, S. (2008). Soundscape characteristics of an environment: A new ecological indicator of ecosystem health. In W. Ji (Ed.), *Wetland and water resource modeling and assessment: A watershed perspective* (pp. 201–211). Boca Raton, FL: CRC Press/Taylor & Francis Group.

Samuels, D. W., Meintjes, L., Ochoa, A. M., & Porcello, T. (2010). Soundscapes: Toward a sounded anthropology. *Annual Review of Anthropology, 39*, 329–345.

Spindler, G., & Spindler, L. (1982). Roger Harker and Schoenhausen: From familiar to strange and back again. In G. Spindler (Ed.), *Doing the ethnography of schooling: Educational anthropology in action* (pp. 20–47). New York: Holt, Rinehart & Winston.

Spradley, J. (1979). *The ethnographic interview*. New York: Holt, Rinehart, & Winston.

Sterne, J. (2003). *The audible past: Cultural origins of sound reproduction*. Durham, NC: Duke University Press.

Truax, B. (1978). *The World Soundscape Project's handbook for acoustic ecology*. Vancouver, BC: A.R.C. Press.

Visweswaran, K. (2010). *Un/common cultures: Racism and the rearticulation of cultural difference*. Durham, NC: Duke University Press.

Wolcott, H. F. (1990). On seeking—and rejecting—validity in qualitative research. In E. W. Eisner & A. Peshkin (Eds.), *Qualitative inquiry in education: The continuing debate* (pp. 121–152). New York: Teachers College Press.

7 Songs to Nowhere

When Critically Creative Processes Meet Impotent Curricular Products

Crystal, Elise, John, and David's chairs are bunched together around the short end of a table built to accommodate two students, while their peers work at their computers and are up at the SMART Board on the far wall of the classroom. Directly in front of the students sits a small music studio that consists of a two-octave MIDI-keyboard, a condenser microphone suspended upside down in a spider web-like shock mount like those seen in music videos, and an old, donated Macintosh desktop computer loaded with the most recent version of Macintosh software and operating system the machine can handle.

Crystal sits in front of the MIDI keyboard draped with a piece of paper with some lyrics the students have agreed upon. All of the other students sit to Crystal's left, first Elise, then John, followed by David who sits on the other side of the microphone stand around the corner of the table.

OpeningVignette.mps3

Mrs. Raymond[1] and the fifth graders in her class were part of a program that brought local students and teachers to the Technology Center for Educational Research (TCER). Located on the campus of a large Midwestern university, classes came through the TCER in six-week stints for three-hour shifts, one class in the morning and a second in the afternoon. As part of their time at the TCER, fifth graders like Crystal, Elise, John, and David were participants in an interpretive research project designed to examine how teachers and their students use processes of music-making to understand academic content. It was this yearlong study that served as a pilot for what became the longitudinal sonic ethnography described in Chapter 6.

Rather than music-making as a genre-independent possibility, the project specifically began as a study of how elementary-aged students used hip-hop to examine and comprehend academic content. At the time, I was working from experiences, personal and professional, a love of music and hip-hop, and what was then a growing literature that hip-hop can serve as a pedagogical vehicle for students to explore academic and sociocultural understandings (Guy, 2004; Hanley, 2007; Richardson, 2006; for literature since 2007, check, for example, Emdin, 2010; Land & Stovall, 2009; Love, 2012).

Bolstered by understandings of a hip-hop audience that is diverse in terms of sociocultural precepts (e.g. race, class, and gender), nationality and culture (e.g. Cepeda, 2006; Chang, 2006; Pardue, 2007), I hypothesized that participating classes would have an interest in hip-hop regardless of the race, class, or gender of their student populations. To these ends, in the fall of 2007, I placed the small music studio in the TCER as described in the opening vignette.

However, as so often happens, through their everyday choices, students casually pushed back at my scholarly assumptions about the ubiquity of hip-hop and their desire to traffic in its potential. Yes, students used the mini studio to rap, create beats, and occasionally spit rhymes about topics related to academic content they were learning at school (as well as topics that were informational but perhaps less typically school-related). They also made songs in R&B, gospel, and rock formats in ways that illustrated the complexities at the intersection of race, class, gender, and music. For example, a pair of African American first grade girls in another class at the TCER sang one of the more musically sophisticated song in terms of melody, timing, and improvisation that any group produced in the study—a song about how they love their mommy and their daddy too that left not a dry eye when parents came out to the TCER to see what their children had done during their time there.

By this I do not mean to imply anything as ridiculous as the possibility that young African American girls cannot create complex, sophisticated music or that I was surprised that they did. Rather, I wish to underscore that movement of this project away from being a study about hip-hop to become an understanding of music-making was the result of students' musical decisions when provided the opportunity to create songs for themselves. In this way, what began as a study of how students and their teachers use hip-hop became an examination of the educational contexts that informed daily classroom interactions, their teachers' curricular and pedagogical decisions, and the curricular meanings that emerged as students wrote their songs about academic content.

This chapter focuses on Mrs. Raymond and the fifth graders she taught, one of the four classrooms observed during this yearlong study. In the tradition of interpretive studies about the ways in which curriculum is enacted in classrooms (e.g., Gilbert, 2013; Mendez, 2008; Page, 1991; Spindler & Hammond, 2006), I consider how teachers and students actively worked within given sets of sociocultural contexts to negotiate curricular understandings. My work here is therefore an examination of both how music-making was used as a curricular tool and the meanings that emerged from its use—an analysis that necessarily includes in its scope questions regarding the ways in which Mrs. Raymond constructed knowledge and the students' processes as they made music together.

As the case of Mrs. Raymond's class documents, the intersection of curriculum, pedagogy, and studenting that is curriculum-in-action (Schwab, 1969, p. 12) is indeed a messy (Lather, 2007), negotiated space. In this instance, Mrs. Raymond *both* actively sought experiences that enriched her students' lives, moving away from the routinized pedagogy and pacing of her regular classroom day, *and* worked to ensure that the curricular meanings lessons provided were akin to "known information questions" (Mehan, 1979) on selected response standardized assessments. In a similar fashion, students worked together in collaborative, critical, and creative ways. They also actively participated in the reconstruction of narrow curricular understandings that were in many ways contradictory to the critically creative collaborative manner in which they worked with one another to create music about that academic content. When combined, this chapter provides a window into the complex, nuanced relationship between formal curriculum, teaching, and students as it is enacted on a daily basis in classrooms.

For these and other such reasons, this study of how music-making can be used as a curricular tool is important because it can serve to shed some light on how active participation in the arts can engender collaboration and critical and creative thought—an understanding that in this instance underscores the importance of students being given the curricular space to construct knowledge with one another. It is also a reminder about the possibilities and limitations of curricular tools. As with all aspects of curriculum, the power and potential of a curricular tool lies not only within the tool itself but also in how it is contextualized and used (Pace & Hemmings, 2006; Page, 1990).

A Brief Note on Culture and Curriculum

Culture and curriculum are contested terms. They are associated as much with the important expression of individual and group's sociocultural norms and values as they are constructions through which (Western, Anglo) dominant groups have asserted their superiority over others under the guise of scholarship (culture) and education (curriculum) (on culture, check Blanchard, Bancel, Boëtsch, Deroo, Lemaire, & Forsdick, 2007; Blanchard, Boëtsch, & Snoep, 2012; Rosaldo, 1993; Visweswaran, 2010; on curriculum, Au, Brown, & Calderón, 2016; Castenell & Pinar, 1993; Malewski, 2009; Pinar, Reynolds, Slattery, & Taubman, 1995). Adding further to their already complicated nature, culture and curriculum are no less impactful or real in spite of their invented nature, maintained and contested through ever-increasing social layers of everyday interactions that run from the immediately local to the international. For these and other such reasons, the constructed and contested nature of culture and curriculum can be helpful in examining what happens in classrooms, a central location for the continuing maintenance of anything that might be called either "curriculum" or "culture" (e.g., Anderson, 2009; Gilbert, 2013; Jackson, 1968; Rist, 1973; Page, 1991).

Because curriculum reflects not a single set of truths but a purposeful set of choices, a study of how classroom actors use curriculum reflects both the kinds of academic content that "count" as knowledge and the local and less local sociocultural norms and values that frame how it is enacted. Curriculum is cultural (e.g., Trueba, Gutherie, & Au, 1981; Tobin, Wu, & Davidson, 1989; Spindler & Hammond, 2006). How then might one consider enacted curriculum, the ways that local classroom actors' negotiations combine to (re)create curricular meanings?

Swidler's (2001) metaphor of a cultural tool kit provides one helpful answer to this question. She suggests that how people use culture is analogous to drawing from a tool kit that contains "implements of varying shapes" that are not necessarily well-fitted, easy to use, or always "do the job" for which they are intended (p. 24). Thus, a person makes sense (Ellsworth, 2005; Howes, 2003) of their relation to others and environments whether or not those constructions are either helpful in interpreting sociocultural contexts or how well such constructions are truly understood.

Interpretive studies of classrooms resonate with Swidler's metaphor. For the past 40 years, such studies have demonstrated how implicit sociocultural norms and values in schools have consistently combined to marginalize students in classrooms based on their race, ethnicity, gender, sexual orientation, language, and/or perceived abilities, be they cognitive or physical (e.g., Gilbert, 2013; Heath, 2006; Valenzuela, Prieto, & Hamilton, 2007; Varenne & McDermott, 1998).

Decisions about academic content, how students used a small recording studio to write songs about content, and the purpose of its use were intentionally left open for teachers to make. In light of Swidler's metaphor, I wondered how teachers and their students would negotiate the use and meanings of these technologies. I was also curious if the process of recording songs would be a useful tool for either teachers and/or students or if it would become yet another misshapen utensil unfit for classroom use. Mrs. Raymond elected to assign students a particular U.S. region to study but left how students addressed that content at the mini studio for the group to enact as they saw fit. This pedagogical decision provided an opportunity for students to express themselves critically (interrogate their processes as individuals and as a group) and creatively (making of music and lyrics) about academic content.

How curriculum is understood by local actors lies at the heart of curriculum studies (He, Schultz, & Schubert, 2015; Kridel, 2010; Malewski, 2009; Pinar, Reynolds, Slattery, & Taubman, 1995). Because the knowledge students receive is always simultaneously academic and social, curricula that are implicit and unintentionally delivered to students are of at least equal importance to the ideas and ideals that are explicitly intended. Capturing these layers of meaning requires a consideration of: a) the intent and expression Mrs. Raymond's formal curricular decisions as they were delivered in the classroom, b) the contexts that informed her decisions, and c) how her curricular choices informed students' use of the mini-studio while at the TCER.

Contexts and Method: Researcher Role, Analysis, Participants, and the Evolution of a Project

The Technology Center for Educational Research is designed to provide students, teachers, and researchers with the latest technologies available with enough computers for up to 25 students to have their own designated laptop or desktop. In the fall of 2007, in addition to computers, there were a multitude of technologies available to teachers and students, such as many handheld video recording devices, a SMART Board, and a camera called "Wolf Vision" that projected pages or any other objects placed in its view onto the room's large screen for all to see. Five ceiling-mounted cameras located throughout the room rotated 180° so that researchers in the adjoining observation room could use them to closely observe all corners of the TCER. The room was also equipped with an additional camera and large screen TV for videoconferencing and microphones were embedded throughout the classroom ceiling that recorded the audio accompanying video recordings, a topic to which I return below.

There was also technical and moral support in the form of three full-time staff members who were available to help teachers with technical and educational matters. As Marge informed me of her role as a TCER staff member, "we're here so that they don't have to worry about the technology but can just use it." According to the TCER's director, the staff's role was to facilitate the use of the classroom and provide support for teachers but was not to extend to teachers' curricular or pedagogical decisions.

Faculty and staff associated with the TCER welcomed my project and the donated hardware. They created the space for the necessary equipment—a donated e-Mac computer so that students could use what was then the most recent version of Garage Band (a music-creation program native to all Macintosh computers), a two-octave MIDI keyboard, microphone, and mic stand—as well as the curricular space for it to be offered to incoming teachers. Garage Band is now well known for allowing the user to construct musical songs rather quickly by dragging and dropping prefabricated music samples; it is also a program of some depth where more sophisticated users can build their own samples, create multi-track recordings, podcast, and otherwise manipulate recorded sounds (vocal or non-vocal).

As I was interested in not only how teachers and students would use this mini studio but also in if they would consider its use, participation in this project was offered to all eight teachers who used the TCER that academic year (2007–2008). This understanding also jived with the way in which TCER staff was to support but not dictate curricular and pedagogical choices.

The summer before they are to spend six weeks in the TCER, teachers attend an initial workshop in which they are introduced to the various technologies available during their stay. It was during this workshop all eight potential teacher-participants were given an opportunity to use the mini-studio, provided with a step-by-step handout of "screen shots" showing each step of the song-creation process that they took home regardless of whether or not they elected to participate in this study, and letters of explanation about the possibility of participating in this study.

Because of the ways in which the facility's use is divided for participating classrooms—two three-hour daily sessions over a six-week period—most of the local teachers and students who use this space throughout the academic year are from local elementary schools. Half of the eight teachers who used the TCER in the 2007–2008 year elected to participate in this project, two fifth grade classrooms, a first grade classroom, and a kindergarten room.

Researcher's Role, Information Gathering and Analysis

My role in this project was twofold, as technological adviser to teachers and as interpretive researcher. Lines between these roles were fluid and I found myself in the classroom working with students more than I had originally intended. This said, teachers made all pedagogical and curricular decisions; I made only technical suggestions to students regarding their music-making. These suggestions were almost solely about how to use the program Garage Band. The few instances in which I worked with students on their lyrics did not apply to Mrs. Raymond or her students but instead to the first graders and kindergarteners who participated in this project.

Each of the four classes elected to use a learner-of-learners model, after a one-day session with each teacher on how to use the mini-studio. My primary role with the two fifth grade classrooms was to teach the first group of students how to use the mini-studio. After that moment, one person from the previous group was to help the next group with the technical information they needed to create their own song. This decision was not only pedagogical but highly practical in nature as I taught two days a week when students were in the TCER, allowing me to work with classes only each Monday, Wednesday, and Friday.

However, this was not what happened in practice in Mrs. Raymond's classroom. Instead, I would arrive in class to find either a) students waving at me through the one-way glass to come in and help the next group use the studio or b) be met upon arrival by either a student or Mrs. Raymond with a request to help the next group who had become stuck in trying to learn the program on their own on days when I was out teaching. This said, I made no decisions about how the mini-studio was to be used, who was to use it when, or any other curricular decisions other than offering the mini-studio as a possibility for teachers to explore while at the TCER.

As briefly noted above, each day's lesson was videotaped on two cameras, one focused on students working at the mini-studio and the other as wide a view as possible of the rest of the classroom in order to record both students' music-making and the classroom interactions that contextualized their interactions. These two cameras and multiple microphones recorded whether or not I was present and was kindly run by TCER staff in my absence.

Over the course of reviewing about 30 hours of audio for this chapter, I came to realize that the sound methods described in the previous chapter were moving forward over the course of this initial yearlong study. Not only had I moved from handing students microphones as I have done throughout my time as a scholar but I had begun finding multiple microphones more helpful than multiple cameras within the first three months of this pilot. This was because audio was geared to record simultaneously with the video and, aside from microphones I placed in the room or others placed in my absence, audio recordings required video. I then either ripped the audio from the video or, in the interest of time, simply minimized the screen I could listen and make notes.

In addition to recordings, I observed Mrs. Raymond's class for the duration of their six weeks in the TCER. Regardless of my presence, at least two cameras and microphones recorded classroom interactions at the mini-studio and elsewhere throughout the room most days the class was in the TCER. Additional information gathered included field notes, documents such as the lyrics sheets students wrote, and interviews with classroom actors that were audio- and, far less frequently, video-taped.

Analysis of the sonic and visual[2] information was an iterative and recursive process in which I tacked back and forth between various pieces of information (recordings, notes, documents, etc.) in order to construct understandings that helped render local actors' interactions sensible to others (Agar, 2004)—a process of continually re-adjusting until a particular understanding explicated idea and ideals within and across information gathered.

Participants in Mrs. Raymond's Class

The group of 16 working-to-middle-class fifth graders represented in this chapter reflected the broader population of House Elementary, a school close enough to campus for students to walk back and forth to the TCER each afternoon. Reflecting the overwhelmingly Anglo county in which the university resides, of the 16 students in the class, 14 were Anglo, two self-identified as African American, 10 were male and six were female. Mrs. Raymond is also Anglo.

Prior to returning to teaching four years ago after raising her three children, Mrs. Raymond taught elementary school for four years in Chicago. Mrs. R. spoke respectfully to her students, consistently addressing them in a level tone and only raising her voice to call the class to attention. She was well-liked by her students, teacher's aide, and the support staff at the TCER; I also found Mrs. R to be a kind and helpful person.

Mrs. R. was made aware of the program at the TCER through her colleagues who had previously participated in the program and was excited at the opportunity to learn more about technology and its possible classroom uses. For reasons detailed later in this paper, Mrs. Raymond decided to use time at the TCER focusing on social studies content and her time at the TCER was designed around the state content standards on "regions of the United States."

Mrs. Raymond's Curricular and Pedagogical Decisions: Curricular Contexts and Teacher Expertise

Teachers are key conduits for the knowledge students receive in classrooms. As such, they are central to curriculum and strongly inform students' classroom roles (Metz, 1978; Pace & Hemmings, 2006; Page, 1991). The curricular decisions teachers make are not only negotiated with students but are also contextualized by local and less-local understandings of what it means to be a teacher and a student (e.g., McNeil, 2000; Valenzuela, Prieto, & Hamilton, 2007; Valli, Croninger, Chambliss, Graber, & Buese, 2009). Situating Mrs. Raymond's decisions about how she utilized the mini-studio and the content she sought to deliver to students therefore requires a consideration of not only her choices but also the contexts that circumscribed those decisions during her time in the TCER.

Contextualizing Mrs. Raymond's Curriculum and Teaching: Visions of Social Studies Education and Teacher Expertise

History-social studies is often presented as a single narrative line of linked dates and events that combine to create a large master narrative rather than a socially constructed, complex composite of multiple perspectives and possibilities (e.g., VanSledright, 2008; Ross, 2010). The National Council for Social Studies (www.socialstudies.org) supports a critical practice of investigating and interrogating existing historical narratives, researching additional possible histories, and the comparison and contrast of a variety of historical perspectives to arrive at a more in-depth, nuanced, and dialogic vision of history-social studies.

Schultz's (2008; Schultz & Oyler, 2006) work with fifth graders in which they "engaged in a seven-month integrated curriculum built around a community-based social action project" (Schultz & Oyler, 2006, p. 423) is an example of this principle in action. Based on inquiry into the meanings of citizenship, democracy, and participation and how those concepts could be put into practice, Schultz created the pedagogical/action space for his students to become advocates for their own cause—the establishment of a new school to replace their own "deplorable" classroom conditions, a school that had been previously promised by the Chicago Board of Education.

However, such cases are increasingly rare, particularly in light of local and federal pressures teachers face to enact curricula as written in order that students might score well on annual standardized assessments of content knowledge (Gershon, 2017; McNeil, 2000; Valli et al., 2009; Valenzuela, Preito, & Hamilton, 2007). As documented in other interpretive studies of elementary social studies schooling (e.g., Heafner & Croce, 2008; Pace, 2008; Wills, 2006, 2008), teaching social studies in elementary schools is strongly informed by the standardized assessments students are mandated to take.

For example, an increased focus on language arts and math in federally mandated standardized assessments can result in local and less local pressures for elementary school teachers to spend as much time as possible on those subjects, greatly limiting the amount of time teachers have to teach social studies (Wills, 2006, 2008). Along similar lines, Pace (2008) suggests that such pressures for ever-increasing annual standardized test scores in language arts and math manifest along class lines. Middle and upper class schools whose students generally score better on their standardized assessments often have the necessary academic cushion to add a less-tested subject like history-social studies to their schedules.

How teachers conceive of teacher expertise is another important context for understanding the ways in which teachers conceptualize curriculum and its delivery in their classrooms (Altbach, Kelly, Petrie, & Weis, 1991; de Castell, Luke, & Luke, 1989). Generally speaking, teacher expertise is constructed in one of two overarching ways. On one hand are scholars who understand teacher expertise is a prescribed set of skills that lie outside teachers themselves and is measurable by their students' success (or lack thereof) at predefined goals and objectives (for example, Floden & Klinzing, 1990). From this perspective, not only could students' scores on annual assessments reflect their knowledge of content but also their teachers' expertise (read competence) at teaching the requisite content as suggested by the federal mandates in NCLB. In contrast, others like Lampert & Clark (1990) situate teacher expertise within teachers. For them, teacher expertise is situational and fluid, based on an understanding that teachers have the necessary expertise to make strong, informed curricular and pedagogical decisions.

As this next section illustrates, Mrs. Raymond's curricular decisions were simultaneously aligned with and complicated more mainstream constructions of social studies content and teacher expertise. Mrs. R. did seek out, create assignments and activities for, and utilize a good amount of available technologies in the relatively unique educational context that was the

TCER. Yet, Mrs. R.'s mainstream understandings of social studies content and teacher expertise created an environment in which students often arrived at narrow curricular meanings that functioned more like responses to items on selected response questions on standardized assessments than deeper explorations of the connections between geography, history, and society endorsed by the NCSS and social studies scholars such as VanSledright (e.g., 2008, 2009) and E. Wayne Ross (e.g., 2008, 2010).

Mrs. Raymond's Talk About Curriculum and an Implicit Construction of Teacher Expertise

Mrs. Raymond's talk about the content she taught during her time in the TCER and how she elected to deliver that content was strongly informed by the kinds of pressures teachers face as described in the previous section. For example, when I asked her about how she decided what content she wanted to teach, Mrs. Raymond's response explicitly connected formal curriculum, the knowledge teachers intend students to learn (Apple, 2014; Page, 1991), to the annual tests her students faced.

RaymondonContent.mp3

In the above description of how she and her colleagues scrambled to put together a social studies curriculum the previous academic year, Mrs. R.'s concerns were not simply that her students would be assessed in yet another subject area, assessments that she emphasized mattered in how Mrs. R. and her students would be understood. Mrs. Raymond's concerns were compounded by the fact that she and her fifth grade colleagues at House Elementary had no textbook with the fifth grade content students needed to pass in their impending assessments.

The pressure she felt was deepened further by differences between her comparatively high scoring class the previous year and the group of children with which she worked the year she was at the TCER.

RaymondonStudents.mp3

The large amount of possible content upon which students might be tested added to Mrs. Raymond's concerns. As she shared with me in another interview: "Science is about eight little benchmarks that need to be met. Easy. Social studies is a full page. It's huge." In addition, Mrs. R. was locally mandated to spend the entire morning on either language arts or "pull outs" such as weekly gym class or trips to the library. As she informed me in another interview, "from 12:10 to 3:10" was all the time "we have to teach social studies, math, science," because of the emphasis on, and large amount of time dedicated to, language arts.

As Mrs. R. noted in her talk above, her unit was indeed a direct response to a specific state standard, identified at the time as Benchmark B.5 for Grade Five Ohio State Academic Standards in Geography: "Explain, by identifying patterns on thematic maps, how physical and human characteristics can be used to define regions in North America." The following details how Mrs. Raymond approached the academic content for students' time at the TCER:

RaymondTCERandStaff.mp3

There is another curricular alignment that is evident both in her talk about putting together a fifth grade social studies book and in this recording about how Marge, a non-academic staff member, was responsible for assembling websites for academic content at the TCER. Whether it is a set of standards that outline a course of study, a book that has the knowledge students' need, or a staff member who assembles websites, curricular expertise lies outside of teachers. Mrs. Raymond is not alone in this belief. Her fifth grade colleagues shared Mrs. Raymond's concern about a lack of a text for students in spite of their respective content area expertise. This would appear to be particularly the case for Mrs. R. who claims a not insignificant background in history and noted her willingness to use that expertise as needed.

Accompanying this construction of curricular expertise is a not-so-implicit understanding of the role and place of teacher expertise. If formal curriculum, the knowledge students are intended to learn, lies outside the domain of teachers and teaching, then teacher expertise is about how teachers deliver that content to students in ways that they themselves can't properly assess. From this position, it is perhaps not surprising that Mrs. Raymond is both worried about her students meeting and, simultaneously, working hard to meet state standards.

Mrs. Raymond's Curricular Decisions and Delivery at the TCER

Mrs. Raymond's decisions about content and its delivery were tightly aligned to the state standards and benchmarks, reflecting her concerns about students' annual standardized assessments in social studies the following spring (2008). An unintended irony of Mrs. R.'s desire to follow the standards in order for her students to be successful on their annual tests was that she utilized the rich resources of the TCER in such a way that all activities functioned not unlike selected-response items found on standardized assessments. As I describe here, this construction of knowledge applied not only to the songs that students created using the mini-studio but also to all of the other activities Mrs. Raymond utilized while at the TCER.

In order to teach her students about the regions of the United States, Mrs. R. created four color-coded teams, purple, green, yellow, and blue, each consisting of four students. After a general information session on each region of the United States, students split up into their groups and conducted further research on the part of the U.S. that had been introduced. Following some time researching, students then dispersed into four different center-like activities in the TCE, half of which I focus on here as illustrations of the curricular function of these centers.

While one group wrote postcards about aspects of the places they found during their research as if they were visiting that particular locale, another group mapped an imaginary journey through their particular region of the U.S. and wrote the names of states on a blank U.S. state map on the SMART Board. A third group created a "clicker quiz" based on the information learned during the general information session about the U.S. region at hand. In this activity, one group of students created selected-response questions about the region that were projected on the large drop-down screen at the front of the room while the remainder of the class used hand-held devices to choose the answer they felt to be correct. The final group used their research to write a stanza of four lines that served as the lyrics for a short song about their regions created at the mini-studio. In addition to these four activities, students wrote weekly emails to Mrs. Raymond that included what she called "five meaningful facts" to which she responded and graded.

> *Clicker quiz. Students are taking a "clicker quiz" written by the four members of the Yellow Team. They sit at their table groups around the TCER with a remote control they call a "clicker" and topographical maps of the United States that do not show state boundaries. Projected on the large drop-down screen at the front of the room is the following question and possible answers:*
>
> *What ocean forms the southern border of Rhode Island?*
> 1. *Pacific Ocean*
> 2. *Gulf of Mexico*
> 3. *Atlantic Ocean*
>
> *Using their clickers, students select the answer they feel is correct. According to the instantly available diagnostic tools associated with this program, 100% of the available participants participated. 69% of participants chose 3. Atlantic Ocean, 19% selected 2. Gulf of Mexico, and the remaining 13% of the class chose 1. Pacific Ocean as the correct answer.*
>
> *"So we have different results again," Mrs. Raymond says as the results post to the screen. "Most interesting. What is the best answer? What ocean forms the southern border of Rhode Island, Candie?"*
>
> *"The Atlantic?" Candie replies*
>
> *"The Atlantic Ocean. Put your finger on the Atlantic Ocean," Mrs. R. instructs as she continues.*

Although the clicker quiz was the activity where students' actions most explicitly mirrored the choices provided on annual state tests, a purposeful curricular decision by Mrs. R. that she used as "a means to get a post [test] so we can see what they learned," all four activities were strongly connected to both standards and testing. For Mrs. R., this connection often manifest in the students arriving upon curricular knowledge that could be understood as measurable, objective facts, as in the following statement: "then I had to make it meaningful. So the meaningful part was finding some five facts about the different [geographic] place they were visiting."

The "facts" that students collected were interchangeable between the four activities and served the same function, an understanding that was explicit in students' music-making. For example, the requisite lyrical content of students' songs was four facts about their assigned regions, each of which became a line that combined to form a single verse for their songs about their particular region of the U.S. Consider, for example, the lyrics to the Yellow Team's s ong:

YellowTeamSong.mp3

This kind of knowledge mirrors the kinds of information students provided in their clicker quizzes, a series of readily identifiable "facts" with single correct answers that could be easily rendered into selected-response items, "known information questions" (Mehan, 1979) with single correct answers.

Additionally, as was the case with the Yellow Team's song, the content songs contained was largely subjective. For example, the number of states in a given region of the United States varies depending upon how one conceptualizes groups of states. The simple addition of a Mountain Region or a Mid-Atlantic States Region would inexorably have altered students' factual statements about their regions. Along similar lines, discussion of the socially constructed and contested nature of statehood for states such as New Mexico, Texas, and California could have changed how student conceptualize either "regions" or "United States" while maintaining the assignment's close alignment to the state benchmark.

Mrs. R's understanding of teacher expertise as generally residing outside of herself was clearly evident during her time in the TCER. Instead of looking to textbooks for content knowledge, Mrs. Raymond turned to others at the TCER. For example, rather than give students a list of websites she compiled or have students do an online search of their own for each region of the Unite States, Mrs. R. asked Marge to compile a set of websites for each state and region. The list that Marge graciously compiled consisted of approximately three or four websites for each state, organized by region. The class used this list of websites to the point that all of the facts that students represented in their four activity centers were from this list, including understandings of which states were in each region of the U.S. Moreover, students only used but one link from each state, the first link listed from a single site that Marge found. "I just copied the next state's name in that part of the URL," she informed me (e.g., Tennessee: Facts, Map, and State Symbols; www.enchantedlearning.com/usa/states/tennessee/).

Although this can be seen as an instance of a teacher using her available resources, a brief mention of how the other fifth grade class that participated in this study similarly covered regions of the U.S. while in the TCER illustrates that this is not necessarily the entire story. Unlike Mrs. R. and her students, this teacher was from an urban school district in which most students were not passing their annual yearly assessments, a factor that placed additional pressure on this teacher for her students to score well on these tests.

However, unlike Mrs. Raymond's class, this group of fifth graders spent their six weeks in groups examining contemporary socio-economic questions and their impact on the people who live there. For example, one group examined questions and concerns about over-fishing in the Northeastern United States and its implications for local fishermen/women, the ecology of the Atlantic, and on the broader U.S. population.

Because Mrs. R. and the students in her class were the first of the two fifth grade groups in the TCER that year, Marge extended the exact same lists of online resources that she presented to Mrs. Raymond. That this other teacher's decisions were made in the same context, year, grade level, with the exact same resources, and in the face of even greater pressures for her students to score well on assessments, demonstrates that choices for greater student involvement and inquiry were possible and plausible within rather similar sets of curricular possibilities and constraints to those Mrs. Raymond faced. It would therefore seem as though Mrs. Raymond's choices to utilize staff in this matter was predicated on more than a simple case of convenience and can be understood as an extension of her understandings about teacher expertise and knowledge.

In sum, Mrs. R's choice to present regions in this fashion is indicative of both the stated pressures she felt to get her students to reproduce the content they would be responsible for knowing on their annual standardized assessments. It similarly reflects her understanding of her role as a teacher. As has been previously noted (Goffman, 1961; Ortner, 2006; Tsing, 1993), even in the face of strong pressures to comply with explicit and/or implicit directives, norms, and values, social actors generally retain some degree of agency to negotiate those constraints. In the case of Mrs. R., her understanding of teacher expertise as lying outside of teachers and her construction of knowledge as measurable skills or facts strongly informed her curricular decisions about both content and its enactment.

Locating Mrs. Raymond's Curricular Understandings and Decisions

It is important to note that Mrs. Raymond's curricular understandings and related decisions to represent knowledge as measurable, observable, and largely objective has a long history in United States schooling (Jackson, 1992; Kliebard, 2004; Schubert, 1986). This curricular tradition, one Kliebard refers to as the social efficiency group, is essentially a factory-model of education: a process-product construction of knowledge through which proper pedagogical techniques yield successful student-products

(Beyer & Liston, 1996; Jackson, 1992; Pinar, Reynolds, Slattery, & Taubman, 1995). From this perspective, knowledge can and should be reduced to discreet, linear, and sequential skills that are measured according to prescribed learning goals. Accordingly, where students' success is understood to be the degree to which they can quantifiably follow or otherwise reproduce prescribed goals and objectives, teachers' success as educators is demonstrated in the percentage of students they can lead to adequately demonstrate their grasp of discreet skills according to those same prescribed, often quantifiable, learning goals and objectives (e.g., Bobbit, 1918; Hunter, 2004/1982; Tyler, 1949).

As the past 40 years of scholarship in curriculum studies document, there are three central concerns with ends-means constructions of knowledge. First, as Kliebard (1975) makes clear in his decisive dissection of Tyler's (1949) "rationale," an ends-means creates a focus on whether or not prescribed goals were met rather than questions about what counts, or should count, as knowledge.

Second, as VanSledright and Ross argue of social studies content, what counts as knowledge is of central importance as such decisions include questions about what knowledge is of value but also whose knowledge is valued (Malewski, 2009; Pinar, Reynolds, Slattery, & Taubman, 1995). This is an important set of questions because there is a history in U.S. schooling of treating students' differences from dominant norms and values as deficits (Valencia, 2012; Valenzuela, Preito, & Hamilton, 2007). From this understanding, the further away a student is from an implicit, dominant Anglo, male, middle class, English speaking, straight, Christian, "able-bodied" norm, the more likely that student's differences will be perceived as deficits, a possibility that often directly impacts long-standing inequalities in American education (e.g., Weis & Fine, 2005; Winfield, 2007).

Finally, although discreet factual information is most certainly a form of knowledge, it is a relatively narrow definition of what curriculum might mean. It generally eschews an understanding of meanings as socially constructed, a possibility among many rather than singularly correct notions, a way of understanding history-social studies that is often antithetical to the inquiry-based pedagogy. As suggested above, this kind of measurably demonstrable knowledge can be described as known information questions (Mehan, 1979) in which the purpose of asking questions is to arrive at single "correct" answers rather than the exploration of many possible answers or related concepts (for a discussion of the origins and concerns with right-ness and correct-ness, Davis & Sumara, 2006, pp. 38–42). Thus, it is a curricular understanding that seems to rather consistently construct student successes at some of their peers' expense and, more generally, removes students' wonder about and deeper construction of knowledge in favor of measurable correct answers.

Complicating Mrs. Raymond's Curricular Choices and Delivery

Despite Mrs. Raymond's talk about the need for a textbook over her own materials and curricular decisions that created activities that functioned like standardized assessments, the content she delivered was not from a textbook. Each of the four central activities in which students engaged while at the TCER provided students an opportunity to gain an understanding of the kinds of social studies content Mrs. R. wished to deliver by exploring different technologies in a round-robin fashion. Furthermore, students were arranged into cooperative groups so that they had others on whom they could rely for help when working on their postcards and email and with whom they could consider the remaining activities (SMART Board journey, clicker quiz, and mini-studio). Mrs. R. also elected to place her students in cooperative groups and assign activities where each group could help one another rather than use assignments that emphasized the need for individual effort and measurements of success.

Similarly, in terms of the mini-studio, Mrs. R. did structure her students' time in the TCER so that everyone had an opportunity to use the mini-studio, and all other centers for that matter, for a few days in order that students had enough time to work on their songs from start to finish. Additionally, other than occasionally sorting out a discussion that turned into an argument or facilitating students' transition to and from the mini-studio, Mrs. Raymond largely gave students the space and time to make the music they wanted to create.

Yet, these curricular choices about content and delivery did not interrupt Mrs. Raymond's understandings of knowledge as discreet measurable facts or of teacher expertise residing outside of teachers in textbooks and experts, creating a curricular context that emphasized aspects of content and teaching that are often placed in binary opposition to one another. On one hand, Mrs. Raymond embraced the TCER's nontraditional, technology-rich classroom environment. She utilized the SMART Board, the clickers, and accompanying software that provided graphic results of scores, email, the Internet, and the mini-studio, creating the space for students to use these resources to explore social studies content. On the other, her constructions of curriculum and teacher expertise circumscribed her pedagogical uses of technology, impacting how students utilized these educational possibilities that resulted in their functioning similar to questions and answers on standardized assessments.

Given that the ability "to define the student's role is one of the teacher's fundamental obligations and prerogatives" (Page, 1991, p. 33), how did the above impact students' use of the mini-studio and the curricular meanings they constructed? It is to this topic that I now turn.

We All Agreed and We All Sang: Students' Collaborative Processes, Emergent Curriculum, and Curricular Meanings

This section focuses on the members of the Green Team, the group whose process is described in the opening vignette. I have elected to focus on this group because their processes were typical for all groups in Mrs. Raymond's class, a similarity that I note through occasional comparisons to other group and group members' talk and actions below.

Student Collaboration, Direction, and Control in Music-Making and Curricular Construction

Making music is a collaborative process and a social construction, a partnership between musicians and with audiences according to sociocultural norms and values about what music means and how it should function (e.g., Berliner, 1994; Coulombe, 2009; Lewis, 2008; Zorn, 2000). While the creation and performance of music is collaborative, it is not necessarily a democratic process. For example, it is often the producer of a given project or the company producing and distributing a "record" that has the greatest say in professional recording situations. Similarly, a single leader in many musical groups often makes the majority of the decisions, including duos.

As the Green Team's interactions in the opening vignette illustrate, students in Mrs. R.'s class most often worked collaboratively and meanings were as emergent as they were prescribed. Throughout this group of students' time at the TCER, they excitedly interjected ideas, discussed possibilities, were critical of the work they compiled, and worked together to create a song that all team members claimed as their own. This did not mean there were not disagreements or misunderstandings. Rather, in most cases everyone in each group actively participated in rejecting, accepting, and proposing musical possibilities that eventually became part of a given team's song. For example, although Crystal wrote the Green Team's lyrics she did not become the team's *de facto* leader. Instead, the rest of the team elected to accept her lyrics so they could more quickly begin to use the mini-studio.

Towards the end of my group interview with the Green Team, we had been discussing the choices Crystal, David, Elise, and John made while at the mini-studio. After students described how their group had elected to each record one line of their song, I asked the Green Team how about their decision-making while working at the mini-studio. As can be heard towards the end of the following interview segment, the conversation devolved into about two minutes of silliness and general goofing about that I cut and the conversation picks back up for a few seconds when I again ask about their decision making process and receive an answer.

GreenTeamInterview.mp3

Similar to their interactions in the opening vignette, the Green Team's talk reflects their shared writing and recording responsibilities. Yet, these were not solely individual endeavors that combined into a collective product. Instead, each individual's decision was evaluated by the group who determined as a whole whether or not their choice worked in service to the song they each had in mind as well as the group's collective understanding of how that song should sound. In the end, they did all did come to a common understanding and participate in processes of music-making.

Emergent Process and Meanings

Although there was little discussion between groups on their processes of song writing, an unspoken template for constructing songs emerged over the course of Mrs. R.'s six weeks at the TCER. The Purple Team was the first group to spend time at the mini-studio and they elected to have a four-line song where each line contained a single selected-response-answer-like statement delivered by a different group member. When the Green Team followed the Purple Team's turn at the mini-studio, they too elected to have each group member say their own line and each line was also written as if it were a selected response answer to an item on a standardized assessment. To this they added stating loudly (shouting, actually) their team's name in unison. The lyrics to their song read as follows:

Green Team

Green Team:	The Southeast!
Crystal:	There is Tennessee, Florida, even Washington, D.C.
Elise:	All the states have different capitals
David:	There are 12 states in the Southeast
John:	They all have different names

As documented with the Yellow Team's lyrics earlier in this piece, this was a songwriting construction that all remaining groups adopted, an explicit manifestation of an implicit, emergent student curricular decision.

Despite the collaborative nature of their process at the mini-studio, the curricular content of students' songs reflected Mrs. R.'s relatively narrow construction of curricular meaning. For example, the Green Team's noting that "There are 12 states" in the Southeast region of the U.S. or the Yellow Team's statement that "Illinois' state grass is the Blue Stem," lacks much information about the region's historical, geographical, social, or economic importance, either to those who live in that region or their relationship with the nation and students' lives.

Perhaps of equal if not greater importance in the case of the Green Team's song was that half of their song could refer to any region of the United States, "All the states have different capitals" and "They all have different names." As no state in the U.S. shares a capital and every state has a different name, this information can be understood as what I have come to think of as superfluous curriculum.

Superfluous curriculum can be understood as the opposite side of the coin from the null curriculum, knowledge that is most noticeable for its absence (Eisner, 1979; Flinders, Noddings, & Thornton, 1986). Superfluous curriculum, on the other hand, is knowledge that while generally understood to be factual in nature is so broadly applicable, such a basic prerequisite for the topic at hand, or so irrelevant to the conversation that its meaning lies not in its content but instead in its function. It tends to function as it does here, substituting relevant knowledge with largely redundant information. As such, like the null curriculum, the superfluous curriculum is a curriculum of negation, removing more relevant for less or irrelevant knowledge.

In this case, although students were actively involved in constructing superfluous curriculum as song, as students are in many ways socialized to gain as much knowledge as they can through the least amount of effort possible (e.g., Alexander, Anderson, & Gallegos, 2005; Foley, 1990), their decision to participate in this way is not necessarily unusual. Here, the negation lay largely with Mrs. Raymond. Instead of asking the Green Team any critical questions about their songs (e.g., Is this your best work?) or for them to include more information in their song that was specific to the region they studied, she praised them for their work and held it up as a good example of how students' songs should function. "Did everyone hear that? Did you see how they had four facts about their region and how they included them in their song? Good job, Green Team!"

Furthermore, in spite of a curriculum of measurable "facts," on more than one occasion students walked away with incorrect information or impressions from their research, understandings that were not corrected by their peers or their teacher. As but one example, when I asked the Green Team during a group interview what they had learned about the western region of the U.S., David told me that, "there are four states in the West" and that the "West is really little." Elise agreed that it was "like really small," a comment Crystal corrected saying that "maybe it was that small on a map," "Well, Texas is kinda' big," John added. Along similar lines, Crystal could not remember whether she was currently writing postcards for the West, the area the class was indeed working on, or the Southwest, the region the class had concluded two days before.

Student's Talk About the Mini-Studio

Despite constant reiteration that the use of the mini-studio was at their teacher's discretion, I was consistently greeted by students who wanted to know when they would have a turn at the studio every time I entered the TCER, including days when I was in and out of the TCER multiple times. Students' talk about their time at the mini-studio also demonstrated this enthusiasm. In the group interviews conducted after each group finished their song, students expressed their pleasure at having an opportunity to make music during class, and every student could recall both the content of their songs and the processes through which they were created.

When asked what they "most like about the [TCER] so far," all members of the Green Team shared that they particularly liked using Garage Band to think about academic content. For example, David offered that music-making was "fun . . . just using the microphone was cool" that he "really liked making a song" and described the song his team made as "awesome!"

so loudly it caused the microphone I used to conduct the interview to feedback. Similarly, Crystal liked that she "could talk into the microphone" and that "everyone got to hear what I said"; John "really liked Garage Band and making songs was totally cool"; and Elise also thought it was "awesome that there was, like, a studio in our class and we could use it!" Their classmates echoed their sentiment. Damian of the Yellow Team wondered "when is Mrs. Raymond going to get all that music stuff" so that he might continue to use it in his back at House Elementary, and Grace of the Red Team told her teammates during their group interview of one another: "Using Garage Band was like the coolest thing we did."

In addition to their enjoyment of music-making, the novelty of being able to write songs in the context of learning as well as the studio itself, students' ability to be in relative control of their own learning also emerged as one of the central facets of their time at the mini-studio that students valued. For example, in response to a question from me about what members of the Blue Team most liked about their time at the TCER, James responded that, "It was totally tight to be able to write our own music, our own song. I mean, what we said was about school but the way we did it, we made it up on our own." This thought is also implicit in the Green Team's talk about their ability to make choices about their song; at no time do they talk about their teacher's or my input.

Thus, despite the relative narrowness and superfluous nature of the curricular content present in their songs, the music-making processes in which students engaged to create those songs was collectively negotiated, emergent, and critically creative. In short, music-making was a curricular tool that engendered meaningful educative processes and for expressing academic content. In light of the information presented in this section, as with any curricular tool, the extent to which it is used to understand academic content rests not only in the potential of the tool but in the ways in which it is utilized in a given context.

Conclusion: Curriculum, Music-Making, and Meaning

Attending to the possible ways in which processes of music-making can function as a curricular tool for academic content, this manuscript also provided a window into the relationship between curriculum, pedagogy, and students. It is a complicated node of examination that suggests that complex, largely democratic student group interactions can lead to relatively shallow curricular understandings and that teacher's curricular and pedagogical decisions can simultaneously utilize more novel approaches to teaching while reifying dominant constructions of both academic content and teaching practices.

As Ortner (2006) and others (e.g., Agar, 1996; Pace & Hemmings, 2006; Valli et al., 2009; Varenne & McDermott, 1998) remind us, local meanings are a result of the intersection between local interactions and the various explicit and implicit contexts that frame those interactions. Although different actors have varying levels of ability to negotiate such constraints and to influence local interactions, because meanings are socially constructed between actors, in most instances, all involved are in some way culpable for the ideas, ideals, concepts, and constructs that are rendered locally meaningful.

In this case, Mrs. Raymond did embrace more novel approaches to teaching. She signed up to have her students participate at the TCER, took the time to learn and utilize most of the available technologies, and created lessons and assignments that conveyed the content she needed to cover in more student-oriented and less traditional ways. While at the TCER students in Mrs. R.'s class gained much of the requisite knowledge she was responsible for delivering, content that conformed to the district curricular pacing plan, consistent with state standards and the standardized assessments on those standards that students would take in the spring. Mrs. R. also provided her students with access to a rich technological environment in which they learned to use both cutting edge hardware and software, some of which like SMART Boards are often designated primarily for teacher use.

However, her understanding of teacher expertise as lying outside of teachers in the hands of textbook creators and other "experts" combined with a relatively narrow test-like vision of knowledge also directly impacted what her students came to know in the TCER. As a result, students tended to use the technologies to which they had access as advanced test-taking tools for known answer questions and the narrowness of their curricular constructions went largely unchallenged.

This complicated construction extended similarly to students. On one hand, as evident in the discussion here of the Green Team, their time at the mini-studio was largely a democratic process of discussion, creativity, and critical inquiry. They investigated the program with one another, were all responsible for aspects of the content they developed, and had many open, direct conversations about the musical content of their songs.

Yet, these discussions did not extend to their song's curricular content. Students did not push one another to find more information about their assigned regions or to connect the information they found to any other ideas they were learning while at the TCER. Instead, the academic content that formed the lyrics of their songs functioned as an operational hurdle to vault on the way to creating their songs, a process that students very much looked forward to and had the additional benefits of reinforcing technological skills and musical experience and knowledge, a topic there is not enough space to examine in detail here.

Thus, the answer to whether or not music-making can be used as a curricular tool that students and teachers find helpful for building knowledge and skills as well as exploring academic content appears to be a resounding yes. Additionally, it appears as though the exploration of how students and teachers use processes of music-making to understand academic content can yield rich data about the nature of curriculum, teaching, and studenting.

Yet these answers should also be tempered with other central questions of curriculum such as: what kind of knowledge is constructed, for whom is such content useful, and what norms and values do these constructions contain? In answer to these kinds of questions, it is important to again note that a curricular tool is as useful as the educational contexts in which it is used and the expectations about the ideas and ideals to which its processes and products might be considered.

In sum, it is indeed possible for making music to simultaneously enable critical, creative, group-oriented processes and reify narrow constructions of meaning that in many ways work against those processes. Similarly, teachers can provide rich frameworks for exploration of ideas that lead to reifying dominant narratives about teaching and knowledge. As Swidler (2001) reminds us, it's not only the access to the tools that counts but also which tools are available to whom, how they are used, and to what ends. Finally, echoing a central theme of curriculum work across decades (e.g., Dewey, 1998/1938; Mendez, 2008; Page, 1991; Schwab, 1969), this chapter documents that, in order to understand what curriculum means, requires an examination of not only teaching and the curricular contexts that inform pedagogical practice but also of teaching- and studenting-in-practice, how local actors render educational contexts sensible.

Notes

1 All names are pseudonyms. The exception is adults who both agreed to have their actual first names used and are now retired, a category that precludes students and Mrs. Raymond. To this end, there are occasional breaks in audio files in order to keep participants' names confidential.

2 There are two reasons for noting ocular information in this chapter. First is to underscore the argument that sound doesn't replace vision but, instead, does things that vision cannot. The second reason was empirical practicality: because I was out of the room two of four days a week, having visual cues about which students were working at the mini studio or the whole class activity in which students were engaged that precluded their time at the mini studio was vital information.

References

Agar, M. (2004). We have met the other and we're all nonlinear: Ethnography as a nonlinear dynamic system. *Complexity 10*(2), 16–24.

Altbach, P. G., Kelly, G. P., Petrie, H. G., & Weis, L. (Eds.). (1991). *Textbooks in American society: Politics, policy and pedagogy*. Albany, NY: State University of New York Press.

Anderson, K. S. (2009). *War or common cause? A critical ethnography of language education policy, race, and cultural citizenship*. Charlotte, NC: Information Age Press.

Apple, M. W. (2014). *Official knowledge: Democratic education in a conservative age* (3rd ed.). New York: R outledge.

Au, W., Brown, A. L., & Calderón, D. (2016). *Reclaiming the multicultural roots of U.S. curriculum: Communities of color and official knowledge in education*. New York: Teachers College Press.

Berliner, P. F. (1994). *Thinking in jazz: The infinite art of improvisation*. Chicago: University of Chicago Press.

Beyer, L. E., & Liston, D. P. (1996). *Curriculum in conflict: Social visions, educational agendas, and progressive school reform*. New York: Teachers College Press.

Blanchard, P., Bancel, N., Boëtsch, G., Deroo, E., Lemaire, S., Forsdick, C., & Bridgeman, T. (Eds.). (2007). *Human zoos: Science and spectacle in the age of empire*. Liverpool, UK: Liverpool University Press.

Blanchard, P., Boëtsch, G., & Snoep, N. J. (Eds.). (2012). *Human zoos: The invention of the savage*. Arles, FR: Actes Sud.

Bobbitt, F. (1918). *The curriculum*. Boston: Houghton Mifflin.

Castell, S. de, Luke, A., & Luke, C. (1989). *Language, authority and criticism: Readings on the school textbook*. London: Falmer Press.

Castenell, L. A. Jr., & Pinar, W. F. (Eds.). (1993). *Understanding curriculum as racial text: Representations of identity and difference in education*. Albany, NY: SUNY Press.

Cepeda, R. (2006). Afro blue: Incanting Yoruba gods in hip-hop's isms. In J. Chang (Ed.), *Total chaos: The art and aesthetics of hip-hop* (pp. 71–77). New York: Basic.

Chang, J. (Ed.). (2006). *Total chaos: The art and aesthetics of hip-hop*. New York: Basic.

Coulombe, R. T. (2009). Improvisation and collectivity: Practical application for research. In W. S. Gershon (Ed.), *The collaborative turn: Working together in qualitative research* (pp. 209–222). Rotterdam, NL: Sense Publishing.

Davis, B., & Sumara, B. (2006). *Complexity and education: Inquiries into learning, teaching, and research*. New York: Routledge.

Dewey, J. (1998/1934). *Art as experience*. New York: Perigree Publishers.

Eisner, E. (1979). *The educational imagination: On the design and evaluation of school programs*. New York: Macmillan.

Ellsworth, E. (2005). *Places of learning: Media, architecture, pedagogy*. New York: Routledge.

Emdin, C. (2010). *Urban science education for the hip-hop generation*. Rotterdam, NL: Sense.

Floden, R. E., & Klinzing, H. G. (1990). What can research on teacher thinking contribute to teacher preparation?: A second opinion. *Educational Researcher 19*(5), 15–20.

Gilbert, A. (2013). *Courageous pedagogy: Enacting critical science education*. Charlotte, NC: Information Age Press.

Goffman, E. (1961). *Asylums: Essays on the social situation of mental patients and other inmates*. New York: Anchor.

Guy, T. C. (2004). Gangsta rap and adult education. In L. Martin & E. Rogers (Eds.), New directions in adult and continuing education (Vol. 101, pp. 43–57). San Francisco: Wiley Publishers.

Hanley, M. S. (Fall, 2007). Old school crossings: Hip-hop in teacher education and beyond. *New Directions for Adult and Continuing Education, 115*, 35–44.

He, M. F., Schultz, B. D., & Schubert, W. H. (2015). *The SAGE guide to curriculum in education*. Thousand Oaks, CA: Sage Press.

Heafner, T., and Croce, E. C. (2008, March 27). "Playing the high-stakes accountability game: Social studies on the sidelines." A paper presented at the American Educational Research Association. New York: March 27, 2008.

Heath, S. B. (2006). *Ways with words: Language, life, and work in communities and classrooms* (2nd ed.). Cambridge, UK: University of Cambridge Press.

Howes, D. (2003). *Sensual relations: Engaging the senses in culture & social theory*. Ann Arbor, MI: University of Michigan Press.

Hunter, M. (2004/1982). *Madeline Hunter's mastery teaching: Increasing effectiveness in elementary and secondary schools*. Thousand Oakes, CA: Corwin Press.

Jackson, P.W. (1968). *Life in classrooms.* New York: Teachers College Press.

Jackson, P.W. (1992). Conceptions of curriculum and curriculum specialists. In P.W. Jackson (Ed.), *Handbook of research on curriculum: A project of the American educational research association* (pp. 3–41). New York: Palgrave Macmillan.

Kliebard, H. M. (1975). Critiques of Tyler. In W. F. Pinar (Ed.), *Curriculum theorizing: The reconceptualists* (pp. 70–81). Berkeley, CA: McCutchean.

Kliebard, H. M. (2004). *The struggle for the American curriculum* (3rd ed.). New York: Routledge.

Kridel, C. (2010). *Encyclopedia of curriculum studies.* Thousand Oaks, CA: Sage Press.

Lampert, M., & Clark, C. M. (1990). Expert knowledge and expert thinking in teaching: A response to Floden and Klinzing. *Educational Researcher, 19*(5), 21–23+42.

Land, R. R., & Stovall, D. O. (Eds.). (2009). Hip hop and social justice. [Special Edition] *Equity and Excellence in Education, 42*(1).

Lather, P. (2007). *Getting lost: Feminist efforts toward a double(d) science.* Albany, NY: State University of New York Press.

Lewis, G. E. (2008). A power stronger than itself: The AACM and American experimental music. Chicago: University of Chicago Press.

Love, B. (2012). *Hip hop's li'l sistas speak: Negotiating hip hop identities and politics in the New South.* New York: Peter Lang.

McNeil, L. (2000). *Contradictions of school reform: The educational costs of standardized testing.* New York: R outledge.

Malewski, E. (Ed.). (2009). *Curriculum studies handbook: The next moment.* New York: Routledge.

Mehan, H. (1979). "What time is it Denise?": Asking known information questions in classroom discourse. *Theory Into Practice, 18*, 285–294.

Mendez, Z.Y. (2008). *General education classrooms: An interpretive study of curriculum in higher education.* S aarbrücken, G DR: VDM Verlag.

Metz, M. H. (1978). *Classrooms and corridors: The crisis of authority in desegregated secondary schools.* Berkeley, CA: University of California Press.

Ortner, S. B. (2006). *Anthropology and social theory: Culture, power, and the acting subject.* Durham, NC: Duke University Press.

Pace, J. L. (2008, March 27). Inequalities in teaching history-social studies under high-stakes accountability in California. A paper presented at the American Educational Research Association. N ew York

Pace, J. L., & Hemmings, A. (Eds.). (2006). *Classroom authority: Theory, research, and practice.* Mahwah, NJ: Lawrence Erlbaum.

Page, R. N. (1990). Games of chance: The lower-track curriculum in a college-preparatory high school. *Curriculum Inquiry, 20*(3), 240–281.

Page, R. N. (1991). *Lower-track classrooms: A curricular and cultural perspective.* New York: Teachers College Press.

Pardue, D. (2007). Hip-hop as pedagogy: A look into "heaven" and "soul" in São Paulo Brazil. *Anthropology Education Quarterly, 80*(3), 673–709.

Pinar, W. F., Reynolds, R., Slattery, P., & Taubman, P. (Eds.). (1995). *Understanding curriculum: An introduction to the study of historical and contemporary curriculum discourses.* New York: Peter Lang.

Richardson, E. (2006). *Hiphop literacies.* New York: Routledge.

Rist, R. (1973). *The urban school: A factory for failure.* Cambridge, MA: MIT Press.

Rosaldo, R. (1993). *Culture and truth: The remaking of social analysis.* Boston: Beacon.

Ross, E. W. (2008). Social studies education. In D. A. Gabbard (Ed.), *Knowledge and power in the global economy: The effects of school reform in a neoliberal/neoconservative age* (pp. 367–376). Mahwah, NJ: Lawrence Erlbaum.

Ross, E. W. (2010). Clockwork: Taylorism and its continuing influence on work and schooling. In E. Heilman (Ed.), *Social studies and diversity teacher education: What we do and why we do it* (pp. 33–37). New York: Routledge.

Schubert, W. H. (1986). *Curriculum: Perspective, paradigm, and possibility*. New York: Palgrave Macmillan.

Schultz, B. D. (2008). *Spectacular things happen along the way: Lessons from an urban classroom*. New York: Teachers College Press.

Schultz, B. D., & Oyler, C. (2006). We make this road as we walk together: Sharing teacher authority in a social action curriculum project. *Curriculum Inquiry*, *36*(4), 423–451.

Schwab, J. J. (1969). The practical: A language for curriculum. *School Review*, *78*(1), 1–23.

Spindler, G., & Hammond, L. (2006). *Innovations in educational ethnography: Theory, methods and results*. Mahwah, NJ: Lawrence Erlbaum.

Swidler, A. (2001). *Talk of love: How culture matters*. Chicago: University of Chicago Press.

Tobin, J., Wu, D., & Davidson, D. (1989). *Preschool in three cultures: Japan, China, and the United States*. New Haven, CT: Yale University Press.

Trueba, H. T., Gutherie, G. P., & Au, K. H. (Eds.). (1981). *Culture and the bilingual classroom: Studies in classroom ethnography*. Rowley, MA: Newbury.

Tsing, A. L. (1993). *In the realm of the diamond queen*. Princeton, NJ: Princeton University Press.

Tyler, R. W. (1949). *Basic principles of curriculum and instruction*. Chicago: University of Chicago Press.

Valencia, R. R. (2012). *The evolution of deficit thinking: Educational thought and practice*. New York: Routledge.

Valenzuela, A., Preito, L., & Hamilton, M. P. (Ed.). (2007). No child left behind (NCLB) and minority youth: What the qualitative evidence suggests. [Special Edition]. *Anthropology and Education Quarterly*, *38*(1).

Valli, L., Croninger, R. G., Chambliss, M. H., Graber, A. O., & Buese, D. (2009). *Test driven: High-stakes accountability in elementary schools*. New York: Teachers College Press.

VanSledright, B. (2008). Narratives of nation-state, historical knowledge, and school history education. *Review of Research in Education*, *32*, 109–146.

VanSledright, B. (2009). Thinking historically. *Journal of Curriculum Studies*, *41*(3), 433–438.

Varenne, H., & McDermott, R. (1998). *Successful failure: The schools America builds*. Boulder, CO: Westview P ress.

Visweswaran, K. (2010). *Un/common cultures: Racism and the rearticulation of cultural difference*. Durham, NC: Duke University Press.

Weis, L., & Fine, M. (Eds.). (2005). *Beyond silenced voices: Class, race and gender in United States schools* (Revised ed.). Albany, NY: State University of New York Press.

Wills, J. S. (2006). Authority, culture, context: Controlling the production of historical knowledge in elementary classrooms. In J. L. Pace & A. Hemmings (Eds.), *Classroom authority: Theory, research, and practice* (pp. 33–62). Mahwah, NJ: Lawrence Erlbaum.

Wills, J. S. (2008, March 27). Inequalities in teaching history-social science under high-stakes accountability in California. A paper presented at the American Educational Research Association. N ew York.

Winfield, A. G. (2007). *Eugenics and education in America: Institutionalized racism and the implications of history, ideology, and memory*. New York: Peter Lang.

Zorn, J. (2000). *Arcana*. New York: Hips Road/Tzadik.

8 Sound Art, Social Justice

Black Lives Matter

It was the summer of 2014. People of African descent, often males, usually adults, were being murdered in real time on video recordings, shared millions of times through the virtual ether. DuBois (1903) and Woodson's (1933) lynching of Black people as public lessons about Black inferiority in society and the inhumanity of Blackness transformed through social media so that any public murder could carry the lessons of fear, hate, and the devaluing of Black, Indigenous, Latinx, and Asian lives. As Carter G. Woodson argued and Boni Wozolek (2015) has again recently reminded us, the educational lynching of people of color in schools, the ways in which curricular norms and values recirculate and reinforce layers of sociocultural ideas and ideals, makes actual lynching both possible and plausible. Lynching has strong implications and histories and I mean them. Even the sharing of explicit evidence of murder, an act of utmost significance in seeking justice, helps to carry the lesson to an ever-widening public: too much melanin in your skin is justification for being murdered. Murdered for listening to police, for changing lanes, for jaywalking, for holding a life-like gun in an open carry state, for walking while Black, for reading while you wait for your child, for standing on the porch.

These murders are ongoing and not new. Indigenous people are murdered at a higher rate than anyone. Trans people of color are killed at almost the same if not a higher rate. Men and women of color are murdered at rates exponentially higher than Anglos when one properly measures total population against the total numbers of reported murders (a number that is always smaller than the total number of murders).

It was the summer of 2014.

Eric Gardener had been suffocated to death. Michael Brown had yet to be shot and killed in Ferguson, though that would happen before the summer was over.

I had pitched a piece to the Bergamo Conference for Curriculum Theory and Classroom Practice as part of a new materialism group that has since become a book (Snaza, Sonu, Truman, & Zaliwska, 2016). In the interim, not only had the uprising in Ferguson gone down in response to Micheal Brown's murder but John Crawford had been murdered for standing while holding a BB gun in a Walmart in Ohio, an open carry state.

To bring this hypocrisy home, here are three of many possible examples for which Anglos have not been shot since I first arrived in Ohio a decade ago. A man with a high-powered rifle walked along through the playground at the back of a local elementary school while school was in session. He was not shot or prosecuted, nor was the school placed on lockdown. A "hunter" who was "lost" walked through the middle school and high school parking lots. School was locked down while the issue was sorted, no charges were pressed, and the superintendent sent out a robocall notifying parents later that evening noting that the person was without incident and that there was no call for alarm. A local high school found that an Anglo male student had brought a handgun to school. The school was not locked down, parents were not notified, and the only reason I know about the event is because a confidential email was sent to teachers from the lead administrator letting them know of the day's events. Like the other two schools, there was nothing to report because everything had been handled; this time not even parents needed know because the handgun was no longer at the school and it was no longer an issue. Yet, John Crawford couldn't stand in a store that sells firearms, hold a BB gun, and talk on the phone without getting murdered. I've been in several Walmart throughout Ohio—people regularly hold guns, point them in random directions, sight down the weapon and dry fire, and carry their purchases throughout the store including in the children's aisles—none of those white men had the police called on them or were murdered.

John Crawford was shot only a few miles from where the Bergamo conference is held.

Each time another Black (male) person was shot, their murder was treated as a singular event on national media outlets. "Yet another Black man has been shot by police," was a pivot from the bigger picture to the particular iteration of a specific murder narrative. Talk about his death. Pictures of his death. Video of his death. Consideration of motive for police and the individual murdered. Rationales, questions, backgrounds, and the like often conflated to a veneer of justification for excessive police force. Yet, almost always a lone event, the telling of a singular murder often at the expense of a broader narrative about police, violence, and race (and gender and indigeneity and queerness and. . .).

John Crawford was murdered minutes from the conference.

I was and am furious, deeply disheartened, and mortified not only at the acts of violence but also at lack of legal convictions in the face of what continues to be what I and many others thought was irrefutable evidence—they were filmed killing people for fucks sake. This was not a naïve shock of someone who was self-deluded/privileged enough to think that racially motivated murders by police were somehow not a reality or that a legal system isn't somehow stacked against poor people, people of color, and their multiple intersections with class, for example, is a category that is inexorably intertwined with race in the United States. Yet, on such violence and aggression continues, murder after murder, acquittal after acquittal, exceptions too far and too few between. And, at the time, when a new murder came to light, it was treated as if it was a singular event rather than a piece to an overwhelmingly obvious narrative.

Words fell short to describe my anger, horror, and overwhelming impotence to affect any modicum of change I sought to enact against this continuing affront.

It was the fall of 2014.

What emerged from my continuing consternation was a means for linking together what had been previously under-acknowledged connections about police murdering men of color. The release of audio to accompany previously released security video footage of John Crawford's murder served as the catalyst for a piece of soundart linking multiple murders together. I spent the next week ripping audio from video of Black and Brown men being shot and often killed by police. I then listened back to these audio recordings over and again, to find commonalities that surfaced in deep listening in ways that might not be as evident in videos.

Representationally, I was torn both about whether enough humanity was present in the work so that it didn't become what might be called "murder porn"—overly focused and purposefully setting the table for others to get off on the tragedy of traditionally marginalized people's aggressively violent deaths—and that my freely improvised playing that reacted to the sonic properties of the piece, the tone as well as the content of talk for example, wouldn't be misinterpreted as somehow musically or ethically inappropriate.

In response to these and other similar questions, I worked to ensure that each audio file of a murder had contextualizing information and was not incidentally edited only to showcase murder. My instrumental response attended to all sounds present, those that hit me as they did, and played that recording back on its own to listen for too many clichés (wailing and screaming sounds overly utilized for example) and against the piece as a whole to better hear whether or not my musical choices jived with the piece as a whole. I also worked to sonically speak to the musicality of voices, the rhythm of cadences, and the timing of reactions—I didn't want to incidentally erase joy and love in the face of violent acts that murdered not only people but those very everyday expressions of personhood. This earlier iteration took four takes and a little over two hours to record the accompanying instrumental track and the entire piece was completed in a week's time working between three and six hours a day. By the end, I was emotionally exhausted and devastated, each iteration triggering experiences of shootings I had been near or to those close to me on top of what was already a difficult work. It is important to again note that, in spite of such difficulties and triggering, my own experiences are in many if not most ways light compared to many people's experiences in the United States, everyday events that are tied to questions of race, economics, gender, sexual orientation, birth nation, and other such sociocultural markers.

Sitting and listening to four murders unfold one on top of the other over the course of 12 minutes is a harrowing, visceral experience. This is especially the case when someone else has the controls making the ability to fast forward impossible, as when a piece of sound art is given the time and attention of a "paper presentation" at a conference. We rarely sit and listen to anything in our contemporary moment. It's a rare experience to sit down, turn on a system or plug in headphones, and have listening be the primary if not sole activity—no scrolling through a phone, texting with friends, having a running conversation, but instead just listening. Slowing down, sitting with others, fully attending to sounds (often music) for the sole purpose of listening is an intimate moment, sharing affect of sound, imagination, ecology, and each other.

Murdering people is supposed to affect us. Yet, in many ways we've become largely numb in our everyday experiences where killing is normalized and an expected daily event. Not that we are at all "okay" with the killing but remain increasingly resigned to its inevitability at a social and cultural level, an understanding that not only further normalizes those escalations within and between communities but also allows, for those with the ongoing privilege to do so, a combination that is always related to questions of race, multiple forms of capital, gender, gender identity and expression, sexual orientation, religion, and language.

Creating soundart where majority Anglo educators might sit and listen to what was a national endemic of race murders by police had the potential to interrupt the normalcy of such killings on one hand and impress on audience members the horror and interconnected nature of these murders on the other. A central part of what gives sonic narratives their power and potential is the ways in which sound sets the scene for our imaginations (e.g., Beck & Grajeda, 2008; Goodman, 2010; Thompson & Biddle, 2013). Because people engage their own understandings and imaginations to create visuals with the sonic, these connections and experiences can be more difficult to distance from one's self, unlike a video where such distance is clear and immediate and where closing one's eyes remains possible.

On a personal level, calling attention to hate, violence, and oppression is something that I take seriously. Less important to the point I seek to make here are the specific personal combinations of histories and experiences that lead me to this positionality; what is significant and germane are how I conceptualize my work and the role of scholars in general and in education in specific how I enact my positionality. Central to one's role as a scholar is an obligation to use one's knowledge and expertise to employ one's available power and position to fight injustice, work with the oppressed against oppressors, and seek to make the world a more accessible, righteous, liberatory place.

As educators and scholars of education, this role is of utmost importance, working to ensure that all children have access to an empowering education and that the information provided to children is inclusive, wide-ranging, thought-inspiring, critical in examination of power, people, and context, and creative in its imagination and expression. To Keaanga-Yamahtta Taylor's (2016) distinctions, not merely equitable but liberatory. Seemingly all contemporary educational tendencies to the contrary, the purpose of education should be the consideration and further enunciation of questions, not the ability to correctly identify answers. A piece of soundart with these tendencies is precisely the kind of scholarship I hoped to engender.

As importantly, this work is scholarship in its critically affective and theoretical stances of another sort as well, a sonic response to what I saw as the potential and problems of new materialisms. The possibilities for new materialisms to interrupt and extend post-structural theory, as well as their associated patriarchies, is important. However, decentering or flattening lines between people and ecology can have the (perhaps) unintended effect of ironically employing affect to avoid injustice. That new materialisms is on the rise, with its pivot to distinctly European forms of science, at the very moment that Black, Indigenous, and otherwise non-Western discourses are, at long last, pressing into mainstream scholarship venues and traditions is also concerning. From a theoretical perspective, this work was also a means to articulate points about the humanness in assemblage and lines of flight to fight injustice in scholars such as Jasbir Puar (2007, 2012) and Alexander Weheliye (2014), both of whom share foundational understandings expressed by Sylvia Wynter (2015) and are echoed in arguments made by scholars like Awad Ibrahim (2014).

In terms of questions about artistry, this was art for social justice, a longstanding tradition in scholarship[1], a possibility supported both by W.E.B. Dubois (1926) and John Dewey (1934) for example. What made this work difficult is that I generally strive to put beautiful things in the world. As I've argued during various paper presentations over the past few years, beautiful does not mean pleasant, easy, or complacent and, in the face of oppression and aggression, intentional acts of beauty are political statements. There is a particular kind of love needed to make impactful music—protests, disruptions, and complications through music can come from a place of beauty, grotesque beauty, inverted beauty, troubling beauty. As a person who regularly plays sounds that can be easily cast as discordant and aggressive by mainstream Western sociocultural sonic norms and values, this may seem to be an odd point to make. However, it is likely that this misunderstanding comes from a conflation of pretty and beauty. It has also been my experience that it is my intention towards beauty that gives my contributions as a musician a particular feeling or vibe, it's affective presence. Which brings me to my point here: making a work that I couldn't personally conceptualize as beautiful interrupted my own aesthetic and understandings.

Making this piece broke something in me. Not my belief in people or some kind of naiveté that racially motived murders by police were either new or infrequent. More something I still can't name yet feel in the pit of my stomach. My commitment to putting this sonic narrative together meant that I watched tens of videos of people being shot and often killed over and over again, yet again as I selected them, then again as I listened to them with my eyes closed, yet again as I ripped the audio from video, listening to the audio for hours as I adjusted the volume and other intricacies, and then immersing myself in those sounds as I responded to them through my horn.

Practically, the sound files were adjusted only in their collection, overlapping playback, panned a bit left and right so audience members could better discern one recording from another, and volume so that they were audible at generally the same level. Due to a scheduling/traveling conflict, the piece was presented at Bergamo in my absence. I called in to the session from the road and a colleague kindly placed a microphone by his phone so that I could speak a bit about the piece to those who had been kind enough to attend our session on new materialisms. I informed the audience of the topic of the piece and its contents, shared that it could be triggering, and that I completely understood if anyone wanted to excuse themselves before it started or felt the need to leave during the playback.

In keeping with my ethical commitments, my purpose was to affect thought and make an aural argument through this piece of critical sonic curriculum. In much the same way, following Frederick Erickson's (2004) argument about not surprising participants with results of a study when engaging in more participatory forms of scholarship, I sought to present an argument that would be difficult but important to hear, not to catch people by surprise in the process. Surprise and disruption are powerful and important tools in scholarship, organizing, and visual and performing arts, tools that I believe in and have employed. However, here it was not my intent. When presenting the piece on a panel at the American Educational Research Association in the spring of 2015, after providing a similar explanation of the piece as I did at Bergamo, I muted the saxophone track and improvised live on soprano as the piece played.

That was then, this is now, the fall of 2016. Our next president has been busy putting in place advisers and heads of state that continue to receive praise from previously ostracized hate groups. Men and women of color, Indigenous people, trans people and other individuals from marginalized groups continue to be shot and often killed by police at an alarming rate. Talk about national registries, for Muslims and scientists so far, spraying swastikas on temples, cemeteries, and parks, and violence against LGBT+, people of color, Indigenous people, Muslims, Jews, and women are all increasing in frequency and intensity. The need to continually reassert, loudly, that #BlackLivesMatter is brought into stark relief in the face of increasing evidence to the contrary that marches forward unabated.

Further and continuing education is decidedly needed and, as has been raised in multiple contexts, the burden of educating Anglos about their racism should not be yet another burden people of color are expected to bear. Dealing with the collapsing of racist resonances across time, space, and sociocultural layers (societal, institutional, interpersonal, individual) that is part and parcel of everyday experiences for people of color in the United States is already too much to ask.

Adding to the mix, education, although it can be less overwhelmingly Anglo than it has been throughout its history, is nonetheless a really, really white academic discipline in which an overwhelmingly white teaching force gets Ph.D.'s and then teaches more white future teachers. Curriculum studies, one of my three home fields, has faced a good deal of internal and external critique for what is in many ways a historical whiteness. However, this is often as much a case of a lack of attention to histories in general and the last 25 years in the field in specific, as it is a reality. For example, as I have argued similarly elsewhere (Gershon, 2016a), Louis A. Castenell and William F. Pinar's (1993) *Understanding Curriculum as Racial Text: Representations of Identity and Difference in Education* was published two years prior to *Understanding Curriculum* (Pinar, Reynolds, Slattery & Taubman, 1995). However, as is the case in many if not the majority of academic fields, curriculum studies remains majority white in person and citation, although this too is changing, though perhaps not rapidly enough.

There are important differences in national conversations about the melanin-based-murders of citizens by police for events that would likely be misdemeanors if crimes at all. The repetitive, patterned nature of this endemic is now taken for granted in mainstream media sources as is increasing attention to similarly racialized murders of women of color, Indigenous people, and trans people of color. As but one example, on July 7, 2016, cnn.com covered the shocking back-to-back murders of Alton Sterling in Baton Rouge, Louisiana, pinned down by police and killed in the street, and Philado Castile, pulled over, shot, and killed in front of his girlfriend and her 4-year old child in Falcon Heights, Minnesota. Where Mr. Sterling's murder was recorded by bystanders on July 5th, Mr. Castile's murder was live streamed by his girlfriend, Diamond Reynolds on the 6th of July. Unlike media attention two years earlier, cnn.com reported both murders as linked to a larger, ongoing, nationwide phenomenon in which hundreds of others had been "killed by U.S. police so far this year" replete with graphs and charts (McKenzie et al., 2016).

What has *not* changed is either the continuing murder of people of color by police or the alarming rate at which such murders occur. When the proliferation of Black suffering is commonplace and social media become tools for contemporary lynching, sonic reiteration of those atrocities has an increased potential for becoming a conduit of othering objectification of suffering of already marginalized people. People using their white (male) privilege to avoid attending to suffering, or, for a more liberalized nuanced version, speak of its horrors and use such acknowledgement to avoid further attention or introspection, also continues unabated.

As but one example, consider the liberal position that if one isn't a person of color, then one shouldn't be part of a broader conversation about race. While meant to be respectful—such a position shows an awareness of how whiteness can take over voices and spaces, and can indeed be an important perspective depending on one's context (at a meeting called for critical scholars of color and those in solidarity for example)—this position is also a means for Anglo (males) to exercise their privilege *not* to engage in conversations about race. As such, this position can be understood as the other side of the rhetorical coin in which white people ask people of color to explain race or other rhetorical exercises that reinscribe the false normalcy of whiteness.

There is also a continuing distancing, a parsing of events that, like the article on cnn.com, at once recognizes individual murders, on rare occasion such as this places them next to one another, and places those police killings in the context of an ongoing national pattern, all without placing multiple murders side-by-side. The cnn.com article is relatively rare in that it sought to emphasize the back-to-back murders of Mr. Sterling and Mr. Castile by the inclusion of videos of both of their murders. It's truly awful and shocking to watch two such murders.[2] It is also the case that the murders of Mr. Sterling and Mr. Castile are most often reported in tandem as a means for documenting the continuing phenomenon of people of color being shot and killed by police, a tool that has become another way to focus on a specific chronologically interrelated events that again limit the scope and breadth of the conversation. Here's a pair of emblematic murders, on to the next.

But it is another thing altogether to have to sit and listen to multiple murders unfold. Because this is what we are delivering to generations of children in the United States, a curriculum of lying, choking, and dying (Gershon, 2016b).

The murders in this work are unchanged. This was a decision I made after doing my due diligence as a scholar in general and in creating this work of arts-based research in particular. More directly, I worked to find and hear as many of the police shootings that occurred between the fall of 2014 and the fall of 2016 that I could find on the Internet. I do so in order to better determine the degree to which the shootings and murders used in this piece documented the range, direction, tone, and tenor of the horrifying effects of shooting and often murdering people of color; not that any iteration is excusable or somehow "invalid." To be abundantly clear, as Fred Moten (Harney & Moten, 2013) incisively articulated in his discussion of Black people's "problematic of coalition" with Anglos: "'The coalition emerges out of the recognition that it's as fucked up for you I don't need your help. I just need you to recognize that this shit is killing you too, however much more softly, you stupid motherfucker, you know?'" (pp. 140–141) Not only does police murdering continually marginalized people diminish everyone's rights and lives, normalizing fields of play for police to murder individuals for otherwise minor infractions underscores the depravity of a state and its relationship to all of its members.

In this instance, these sounds represent the police shootings of John Crawford, Michael Brown, Kaijeme Powell, Martin Hall, and Levar Jones. Mr. Crawford was killed for holding a BB gun in Walmart as described above; Mr. Brown, shot with his hands up for yet undetermined reasons beyond police aggression and fear; Mr. Powell was killed for talking back to police after stealing two sodas from a convenience mart; and Mr. Hall, a mentally ill homeless man holding a pen knife, was shot 46 times by eight Saginaw, Michigan police officers. Only Levar Jones survived being shot in the leg after complying with Sean Groubert's request to see his license.

What is updated in this are the narrative arc of this piece and my instrumental response. While the beginning of the piece still emerges out of silence, the silence stretches a bit before sounds emerge in an artistic nod to the ways that police shootings emerge out of the everyday, seem unending and inevitable, a combination that contributes to rising tensions because of the false calm before the next shooting is, to date, always false and helps to reassert the horrific and dehumanizing normalcy that people of color will be shot and killed by police. Adding further insult to injury, both literal and metaphoric, the acts that caused all the African American men in this piece to be shot are situations that untold Anglo children and adults have escaped with a verbal warning: taking a can or two of soda from a store, getting one's wallet when ordered to by police, or raising one's hands in response to police request to freeze.

The work now ends on the less hopeful note of reporting on Mr. Hall's murder rather than with a more hopeful note of protesters singing a version "Which Side Are You On?" during intermission of a concert by the St. Louis symphony, a move during which they unfurled signs and sang verses about "Justice for Mike Brown." All sounds on the associated sound file have been edited for volume to be more audible and the recordings of police shootings occur in the following fashion. The piece begins with the initial call to police that led to John Crawford's murder panned to the left of the stereo field. This recording is joined by reporting on Michael Brown's murder on the right side of the stereo field approximately 2:44 into the recording. As the police recording continues on the left, the sounds surrounding the murder of Kajeme Powell follow the Michael Brown recordings on the right (~5:44) and is joined by the recording of Mr. Jones from the left after the recording of incidents surrounding Mr. Crawford's murder come to a close. The piece ends with reporting and recording of Mr. Hall being shot over 40 times.

All of the sound effects from the baritone saxophone in this recording are analog, the result of combinations of traditional and extended techniques that allow for expressions as multiphonics, the simultaneous sounding of multiple notes in an overtone series, and altissimo register playing, notes in and out of a tempered scale played beyond the keyed ranges of a wind instrument. The lone exception to this is some slight sound sculpting, a hint of reverb and a slight low end push so that the horn sounds as live as it did in the room. This is standard engineering practice that seeks to balance the required false and clean dryness of a recording context. By comparison, nearly every recording you have ever heard has had at least some degree of analog or digital pre or post production and usually with a much heavier hand than I used in this instance.

In spite of all the words in this chapter, and throughout the book, they are inadequate to express the broken, sunken anguish in response to these murders. In lieu of a tenor saxophone, I have elected to use a baritone saxophone in this version, in part due to its cello-like tonality, and in part because of internal affect. Holding a big vibrating metal piece of sound production does something in ways that are phenomenologically significant to this project, moving my own body as it moves air as it moves with the rest of the organized sounds. Improvising, responding, reverberating resonances.

SoundArtSocialJustice.mp3

Notes

1 For example, Steve Reich created a tape-loop piece of sound art called "Come Out" that layered and looped David Hamm's recounting of his beating at the hands of police in Harlem. For more on this story, check: Beta, 2016.

2 To date, Mother Jones' online piece "Here Are 13 Killings by Police Captured on Video in the Past Year" (Vicens and Lee, 2015) is one of the very few instances in which a media outlet has placed multiple murders by police together in a single piece of journalism.

References

Beck, J., & Grajeda, T. (2008). *Lowering the book: Critical studies in film sound*. Urbana, IL: University of Illinois Press.

Beta, A. (April 28, 2016). Blood and echoes: The story of Come Out, Steve Reich's civil rights era masterpiece. Retrieved from: http://pitchfork.com/features/article/9886-blood-and-echoes-the-story-of-come-out-steve-reichs-civil-rights-era-masterpiece/.

Castenell, L. A. Jr., & Pinar, W. F. (Eds.). (1993). *Understanding curriculum as racial text: Representations of identity and difference in education*. Albany, NY: SUNY Press.

Dewey, J. (1934). *Art as experience*. New York: Perigree.

DuBois, W. E. B. (1903). *The souls of Black folk*. Chicago: McClurgand.

DuBois, W. E. B. (1926). *Criteria of negro art*. Retrieved from: www.webdubois.org/db CriteriaNArt.html

Erickson, F. (2004). Talk and social theory: Ecologies of speaking and learning in everyday life. Malden, MA: Polity Press.

Gershon, W. S. (2016a). The sound of silence: The material consequences of scholarship. In N. Snaza, S. E. Truman, D. Sonu & Z. Zowiska (Eds.), *Pedagogical matters: New materialisms and curriculum studies* (pp. 75–89). New York: Peter Lang.

Gershon, W. S. (2016b). Getting schooled: A curriculum of lying, choking, and dying. In B. Wozolek (Ed.), *Special edition of AERA's division B newsletter: Black lives matter* (pp. 17–20). Washington, DC: AERA.

Goodman, S. (2010). *Sonic warfare: Sound, affect, and the ecology of fear*. Cambridge, MA: MIT Press.

Harney, S., & Moten, F. (2013). *The undercommons: Fugitive planning & Black study*. New York: Minor Compositions.

Ibrahim, A. (2014). *The rhizome of Blackness: A critical ethnography of hip-hop culture, language, identity, and the politics of becoming*. New York: Peter Lang.

McKenzie, S., Said-Moorhouse, L., Greene, R., Berlinger, J., Jones, S., Grinberg, E., Shoichet, C. E., & Almasy, S. (2016, July 7). *Two police shootings, two videos, two black men dead*. Retrieved from: www.cnn.com/2016/07/07/us/shootings-alton-sterling-philando-castile/

Pinar, W. F., Reynolds, R., Slattery, P., & Taubman, P. (Eds.). (1995). *Understanding curriculum: An introduction to the study of historical and contemporary curriculum discourses*. New York: Peter Lang.

Puar, J. K. (2007). *Terrorist assemblages: Homonationalism in queer times*. Durham, NC: Duke University Press.

Puar, J. K. (2012). "I would rather be a cyborg than a goddess": Becoming-intersectional in assemblage theory. *Philosophia*, *2*(1), 49–66.

Snaza, N., Sonu, D., Truman, S., & Zaliwska, Z. (Eds.). (2016). *Pedagogical matters: New materialisms and curriculum studies*. New York: Peter Lang.

Taylor, K-Y. (2016). *From #BlackLivesMatter to black liberation*. Chicago: Haymarket Books.

Thompson, M., & Biddle, I. (2013). *Sound, music, affect: Theorizing sonic experience*. London: Bloomsbury.

Valente, J. M., Bahan, B., & Bauman, H-D. (2011). Sensory politics and the cochlear implant debates (245–258). In Paludneviciene, R., & Leigh, I. (Eds.). *Cochlear implants: Evolving perspectives*. Washington, D. C.: Gallaudet University Press.

Vicens, A. J., & Lee, J. (2015, May 20). Here are 13 killings by police captured on video in the past year. *Mother Jones*. Retrieved from www.motherjones.com/politics/2015/05/police-shootings-caught-on-tape-video

Weheliye, A. G. (2014). *Habeas viscus: Racializing assemblages, biopolitics, and Black feminist theories of the human*. Durham, NC: Duke University Press.

Woodson, C. G. (1933). *The mis-education of the Negro*. Complete text online retrieved from www.historyisaweapon.com/defcon1/misedne.html

Wozolek, B. W. (2015). *The presence of absence: The negotiation of space and place for students of color at a predominantly white suburban high school*. Dissertation. Kent State University.

Wynter, S., & McKittrick, K. (Ed.).(2015). *Sylvia Wynter: On being human as praxis*. Durham, NC: Duke University Press.

9 Becoming Attuned

Educational Deep Listening

Over the course of her remarkable career, from her work with the San Francisco Tape Music Center in the 1960s to her work in and through deep listening, Pauline Oliveros was a ground breaking musical and philosophical innovator. She was one of those rare people who was known as much for her compositions as for her freely improvised expressions and as much for her work with emerging media (tape, electronics) as for her virtuosic facility on an accordion, an especially designed instrument that could be played in two different scales of just intonation rather than a more typical Western tempered scale.[1] Oliveros' lifelong commitment to feminism, social justice, sonic attention, and awareness is perhaps best expressed in what she called "deep listening," an approach to music and life that was at the center of her career and everyday experiences.

Of her multitude of accomplishments, one of Oliveros' best known is a series of Sonic Meditations published in 1974 that are at once explicitly feminist, genre busting, and caring. As she states in the second introduction:

Each Sonic Meditation is a special procedure for the following:

1. Actively making sounds
2. Actively imagining sounds
3. Listening to present sounds
4. Remembering sounds

(Oliveros, 1974, Introduction II, para. 4)

As Steve Smith (2016) noted in his beautiful obituary on the occasion of her death, "'Native,' the most commonly cited example, is also the most succinct: 'Take a walk at night. Walk so silently that the bottoms of your feet become ears.'"

Like her Meditation, The Flaming Indian, the anachronistic terminology can call into question possibilities of appropriation and othering. Yet, in 1974, to specifically call for meditations that intentionally speak to Indigenous and other non-Western ways of beingknowingdoing, as she does when calling for "a Japanese bowl gong" (Sonic Meditation, Bowl Gong), can also be understood as an explicit acknowledgement of music from around the world without privileging Western music in general or Western art music in specific. Oliveros' choices can similarly be conceptualized as a disruption of a construction of Western music and their patriarchal attitudes about the superiority of Western art music over all other forms of sonic expression.

Throughout this book, I have often returned to Pauline Oliveros' (2005) notions of deep listening. In her piece, Quantum Listening,[2] Oliveros (1999) notes that, "Deep Listening is listening in every possible way to everything possible to hear no matter what you are doing. Such intense listening includes the sounds of daily life, of nature, or one's own thoughts as well as musical sounds." It is an active process that shapes local and less local sociocultural understandings and is shaped by them, an act of interpretation and choice within a sea of ever-unfolding improvisations that are the human experience, from the most everyday to the least expected and across layers of scale and collapse notions of scale.

Educators could learn a great deal from engaging Oliveros' deep listening practices.

Rather than produce my own educational sonic meditations—as Oliveros' are not field, culture, or individual dependent and are therefore immediately applicable to education—or anything that might be called a manifesto, my purpose here is to present some additional ways that listening deeply to children and youth as daily educational practices could deepen what education could mean and how it might function. This notion also speaks to contemporary understandings about how to approach any project that might be called socially just or in solidarity: by listening. Although written in a directive style in homage to Oliveros' meditations, they are intended as suggestions that are springboards for further refinement, consideration, alteration, or rejection all together.

Start educational interactions in doubt and uncertainty, listening deeply. Because each hearing is a mishearing and what you heard, you misheard. Try to listen again more deeply. Did you hear it again? Always already misheard. Doubt and uncertainty are perhaps two of the deepest educational tools at an educator's disposal. Doubting what you heard will more likely cause you to reflect again in increasing depth and from multiple perspectives, a motion that is reflexive, the continual checking of your own intentions and their expressions. This, in turn, often leads to criticality, in that doubting what you heard and a pattern of checking what you yourself have said is likely to engender the same kinds of doubt and uncertainty about what you think you heard from outside of yourself. This combination of taking purposeful steps to attend to others and how you impact them is central to caring processes, about that which lies outside yourself and to yourself. It is important to note that this is not a self-deprecating, aggressive, or oppressive doubt and/or uncertainty but is instead a means for opening pathways to better and more deeply listening to others and yourself. Using doubt and uncertainty as tools is not the same thing as never having confidence or being relatively certain of things, it is the possibility that even that which you are most confident about knowing may yet be wrong or harmful and, in having that possibility, increase the chances doing less harm while gaining more information in an iterative and recursive fashion. Reflexivity, criticality, and care are central to all ethical educational interactions.

What does wonder sound like? Would you know it if you heard?

Listen to others so hard that you can hear the pathways they travel and the possibilities those journeys might engender. Can you hear where they're going? Will you listen along the way?

What Does It Mean to Listen Deeply to Students?

At its surface, this question raises other associated questions that strongly inform possible answers to the initial question. For example, it raises the question, who is a student, as well as the relationship to another as a student, an orientation that implies the listener is a teacher. Yet, might not all involved be students together? Engaged in forms of study where one person generally has greater experience and understandings than the other (due to her studying over years), but this does not mean that those with less experience or understandings cannot add to the more experienced student's being-knowingdoing. What happens when teachers listen like students regardless of those involved or the location of those interactions?

Walk into a space where you teach (kitchen, hiking trail, classroom, etc.) when no one else is there and sit in a place you might consider "the middle" in a chair or on the ground. Close your eyes and listen to your breathing until you are calm and focused. Then, slowly, bit-by-bit, in small increments, listen outward in an omnidirectional fashion, attending to the sounds that surround you in successive layers away from your seat so that each layer sets and another is added. When you reach the physical walls of the room, or the things that populate whatever ecology you have chosen, listen past them as far as you can. Then, when you're ready, slowly listen back in to through the layers, letting each one go as you concentrate on the next until you end up again focused on your own breathing. Repeat the exercise in reverse, listening as far as you can to as much as you can then narrowing back to yourself before ending as far away as yourself and listening to as much as you can. All the while, listen for what is present in the absence of those things that can and do make sounds, to their resonances and reverberations that still affect you and you are still affecting through your listening.

For Educators Who Work in Classrooms, Regardless of Their Location

Take a moment from listening to the affective feelings of a student-focused activity (small group, whole group, individual, etc., but one in which you are not leading instruction)—the sharp pitches that signal disagreements, the kinds of dull silences of students who are disengaged as opposed to the percussive near-silent slaps of students' frustration, and the rolling sounds of people working—away from the thousands of sounds that have become a natural part of your anticipated soundscape, and tune in as you would to your favorite music. Not just to the prosody of talk but the ways in which the pacing and tone of words, writing implements, moving bodies, whirring machines, and all the other sounds in the room create their own unique sound field, a composite that is as liminal as it is lovely and as mutable as it is consistent in aspects of its articulations.

What is the difference between listening to P–12 students as students and listening to them as children, or listening to them as people or individuals? How can you listen deeply to students if you don't hear them as children or as people?

Who listens and who has the privilege not to?

If you can't hear the complexities of a person in both how they identify themselves and the identifications put on them by multiple iterations of others, how can you deeply listen to her?

Deep listening resides in the ever-evolving emergent now, soundworlds that carry past resonances. Deeply listening to others allows you to ride that curling sonic wave as it unfurls ever in the moment. It takes energy and time and practice. It's a habit of all parts, of sensation and signification, affecting and being affected. It also often means not talking as much.

Ecological Listening

When I think beyond soundscape to the multiplicity of acoustemology and then apply contemporary conversations about assemblages in which things are recognized for the agency they already have—the chair I'm sitting on acting on me as much as I am acting on the chair—I get to thinking about sonic or sound ecologies. The multitude of things that sound and resound within a given space, a space that is as often inclusive of far away sounds not unlike getting a radio station from three states away in a frequency that is often empty one rainy night. A mobile, fluid ecology, ever-shifting and reshaping as things move about, including people. What might it mean to deeply listen to ecologies?

Listening Rights

Everyone has the right to listen and everyone has the right to be heard. Like the ears we possess, whose absence forms a null curriculum of presence in a normalized hearing world, whether hearing or profoundly not, is a human right. For deeply listening to another is as much, if not more, about the intention of an expression and another's attention to that intention regardless of its means of expression, as it is about the perception of sounds-as-sounds. Whether or not one hears the sonic or feels them haptically, or imagines the sounds one once heard or have never heard before, they are nonetheless sonic affects that are affecting. At an interpersonal level, deep listening is becoming attuned to another's expressions so that you can hear not only their intentions but also the nested layers of local and less local intentions, attentions, and expressions that inform the immediate iteration of deeply listening to another. It is the difference between the absurdity of "listening right" as a measurement of correctness and a listening right, the right for someone to be heard as they intend, within and through the inescapability of mishearing every thing, every time.

Listening deeply to individuals, groups, ecologies, and things means fully hearing as much of pieces and their combinations as one can, a process that is furthered by constant practice and awareness of practices. From this position, there is no excuse for not attending to the multiple, complex combinations embedded in people and their experiences. Fully listening to iterations of misheard information insists that there is always more to be heard and that such an act, though significant, cannot stand in for the infinite variety of possible understandings and therefore cannot be an all-curing act of its own. Forms of deep listening, however, can be a means for pushing back at the forking paths of reductionism and essentialism. Both become impossible when faced with the complex intricacies that are audible when working to attune to the multiplicity of overlapping ecologies that move through one's self as much as they surround you, crashing, reverberating waves of resonance. When any individual thing cannot be cleanly separated from other things, an understanding that is in many ways a first step in deep listening, then the ability to reduce any individual thing to its constituent parts or believing to fully understand any-thing as something that might be called whole, becomes what it always was already, impossible.

From these perspectives, expertise is being transparent and clear about what one believes one understands, and the limits of those understandings. One can spend a lifetime listening deeply to a specific set of possibilities and will never fully know them. Doubt and uncertainty again, springboards for inquiry and wonder.

When was the last time you just sat and listened?

Notes

1 An equally tempered scale is the combination of intervals between notes that creates the sounds most common to Western-derived music, from Western art music to jazz to country. It is the tempered scale that gives us the combination of whole and half steps producing black and white notes on a piano, the do-re-mi sounds that produce intervals like an octave or a fifth, and our constructed perceptions of consonance and dissonance. Unlike equal temperament, just intonation is produced by a series of intervals expressed as fractions that are ratios between the scale's fundamental or reference note and another note. This creates two different compositional "problems" that an equally tempered scale "fixes." First, because the ratios are in relation to a reference note (root note), there are potentially an infinite number of possible notes between the reference

note (1/1) and its octave (2/1). Second, each scale functions as its own sonic universe where scales with a reference note of "C" would be different scale if it had a reference note of "D"—while some notes might align, others would not. Pauline Oliveros played an accordion in just intonation and improvised with musicians who played instruments in a variety of tunings that included but was by no means limited to equal temperament (for example, a koto has a wide variety of traditional Japanese tunings).

2 This work is grown from the seed of her keynote address, Quantum Listening (Oliveros, 1999) presented at the seminal conference Improvising Across Borders which "was conceived by Dana Reason, then an -minded graduate student in the Department of Music at the University of California, San Diego (UCSD), and produced in collaboration with her fellow graduate students Michael Dessen and Jason Robinson" (Lewis & Piekut, 2016, p. xi). Reason is a remarkable musician, composer, and improviser in her own right who played in various incarnations with Pauline Oliveros including a trio along innovation with *shakuhatchi* player Philip Gelb called The Space Between, a group that produced three albums between 2001-2003 each with a different guest bassist, Barre Phillips, Matthew Sperry, and Jöelle Léandre.

3 In the context of the email, it's unclear to me if she meant "here" or was referring to a comment earlier in the email so have left it as ambiguously as I found it.

References

Ayers, W., & Miller, J. (1997). *A light in dark times: Maxine Greene and the unfinished conversation*. New York: Teachers College Press.

Lewis, G. E., & Piekut, B. (2016). *The Oxford handbook of critical improvisation studies, volume 1 & volume* 2. Oxford: Oxford University Press.

Oliveros, P. (1974). *Sonic meditations*. Akron, OH: Smith Publications.

Oliveros, P. (1999). Quantum listening: From practice to theory (to practice practice). *Musicworks*, *76*, 37–45.

Oliveros, P. (2005). *Deep listening: A composer's sound practice*. Lincoln, NE: Deep Listening Publications.

Smith, S. (2016, November 27). Pauline Oliveros, composer who championed "deep listening," dies at 84. *The New York Times*. Retrieved from www.nytimes.com/2016/11/27/arts/music/pauline-oliveros-composer-who-championed-deep-listening-dies-at-84.html?_r=0

The Space Between (2001a). *The space between with Barre Phillips* [Compact Disc]. 482 Music.

The Space Between (2001b). *The space between with Matthew Sperry* [Compact Disc]. 482 Music.

The Space Between (2001c). *The space between with Jöelle Léandre* [Compact Disc]. 482 Music.

Index

One of my favorite delights has always been the hidden tracks buried at the end of a recording, suddenly emerging from what had been the quiet hush of nothingness after the echoes of the last song had well since faded. It is perhaps for this reason that I also have really dug Ted T. Aoki's literal and metaphorical use of a "lingering note," something that I may have over-employed already in my comparatively short academic career. It is in this spirit that I'd like to share this brief story, a single interaction of encouragement that in many ways gave me the internal fortitude to move forward from conceptualizations more grounded in music to considerations of the sonic.

Through a series of events, the first article I wrote on collective improvisation as a lens for conceptualizing classroom interactions won an award, something that was as thrilling as it quickly became daunting. What to do as a follow-up? What if that was as good as it would ever get? What if that was the last even remotely smarty-pants thing I had to say.

As so often happens with scholarship, this is, after all, how we end up with the (altogether false) notion that ideas proceed in organized lines of inquiry, I had the idea that I might be able to do the same thing with all of the non-speaking sounds of a classroom interaction. With this in mind, I again set out to painstakingly transcribe and chart all the classroom sounds I could into rows of Excel like a conductor's score, making an argument that one could know that a classroom was in transition without talk.

The scraping of the chairs, and the whooshing of turning pages or thunk of closing textbook covers, as well as the immediate crescendo of activity during the putting-away phase, followed by a rapid decrescendo of most sounds that teachers across the nation take to be elementary-school-leaving-the-classroom-readiness. After sending the piece out to multiple journals, where my feedback was the positive news that an editor found the work of interest and the disappointing news that it didn't fit the parameters of their journal, I sent the piece out to a journal that rejected the work on the grounds that I hadn't read or cited central works in their field. The editors of this last journal were absolutely correct that I hadn't delved into those who had done the work in their area but, after catching up, I realized that the editors of that journal too had missed what I was after.

It was at this critical juncture I had an encounter with Maxine Greene at a conference in New York. She had been the keynote speaker and I, like so many others, queued up to shake her hand and exchange a few words. When it was my turn to speak with her, line still growing behind me, I let Dr. Greene that her scholarship had deeply informed my own including a current line of thought about sounds and improvisation that resulted in a piece I was currently working on. Really, she said, send it to me. And so I did. After some confusion about just who was sending the email that a colleague kindly helped me sort, here was the portion of her reply that spoke to my work, copied and pasted below exactly as it appeared in my inbox that day[3]:

> Your aural notions won't help her; but they are lovely, illiminating notions. I thought I would not understand the paper, but I think I think

> I mostly do.In a workshop here today, we were talking about TO THE LIGHTHOUSE; and I thought of your work in connection with the central section on the dinner party: the voices, loud and whispery; the clink of china; the sound of the door opening and closing,footsteps, outside wind from water Am I on yourbeam or not? If not, please say. I do know it is about class rooms and erasers, et al. Anyway, let's stay in touch. My best, Maxine

And that is how, in a single email, Maxine Greene, in what was but a brief exchange, helped me not feel so lost or odd and like I may have been on to something, as she did for so many over the years (e.g., Miller and Ayers, 1997). It is also but one example of what happens when educators spend a lifetime of listening deeply to others, within, across, between, and around perceived boundaries, and what it feels like when you are truly heard.

For Product Safety Concerns and Information please contact our EU representative GPSR@taylorandfrancis.com
Taylor & Francis Verlag GmbH, Kaufingerstraße 24, 80331 München, Germany

www.ingramcontent.com/pod-product-compliance
Ingram Content Group UK Ltd.
Pitfield, Milton Keynes, MK11 3LW, UK
UKHW022000180926
475701UK00003B/110

* 9 7 8 0 3 6 7 3 4 2 6 0 9 *